From Signal to Symbol

Life and Mind: Philosophical Issues in Biology and Psychology

Kim Sterelny and Robert A. Wilson, Series Editors

From Signal to Symbol: The Evolution of Language, Ronald J. Planer and Kim Sterelny, 2021

What's Left of Human Nature? A Post-Essentialist, Pluralist, and Interactive Account of a Contested Concept, Maria Kronfeldner, 2018

Rock, Bone, and Ruin: An Optimist's Guide to the Historical Sciences, Adrian Currie, 2018

A Mark of the Mental: In Defense of Informational Teleosemantics, Karen Neander, 2017

Mental Time Travel: Episodic Memory and Our Knowledge of the Personal Past, Kourken Michaelian, 2016

Becoming Human: The Ontogenesis, Metaphysics, and Expression of Human Emotionality, Jennifer Greenwood, 2015

The Measure of Madness: Philosophy of Mind and Cognitive Neuropsychiatry, Philip Gerrans, 2014

Beyond Versus: The Struggle to Understand the Interaction of Nature and Nurture, James Tabery, 2014

Investigating the Psychological World: Scientific Method in the Behavioral Sciences, Brian D. Haig, 2014

Evolution in Four Dimensions: Genetic, Epigenetic, Behavioral, and Symbolic Variation in the History of Life, revised edition, Eva Jablonka and Marion J. Lamb, 2014

Cooperation and Its Evolution, Kim Sterelny, Richard Joyce, Brett Calcott, and Ben Fraser, editors, 2013

Ingenious Genes: How Gene Regulation Networks Evolve to Control Development, Roger Sansom, 2011

Yuck! The Nature and Moral Significance of Disgust, Daniel Kelly, 2011

Laws, Mind, and Free Will, Steven Horst, 2011

Perplexities of Consciousness, Eric Schwitzgebel, 2011

Humanity's End: Why We Should Reject Radical Enhancement, Nicholas Agar, 2010

Color Ontology and Color Science, Jonathan Cohen and Mohan Matthen, editors, 2010

The Extended Mind, Richard Menary, editor, 2010

The Native Mind and the Cultural Construction of Nature, Scott Atran and Douglas Medin, 2008

Describing Inner Experience? Proponent Meets Skeptic, Russell T. Hurlburt and Eric Schwitzgebel, 2007

Evolutionary Psychology as Maladapted Psychology, Robert C. Richardson, 2007

The Evolution of Morality, Richard Joyce, 2006

Evolution in Four Dimensions: Genetic, Epigenetic, Behavioral, and Symbolic Variation in the History of Life, Eva Jablonka and Marion J. Lamb, 2005

Molecular Models of Life: Philosophical Papers on Molecular Biology, Sahotra Sarkar, 2005

The Mind Incarnate, Lawrence A. Shapiro, 2004

Organisms and Artifacts: Design in Nature and Elsewhere, Tim Lewens, 2004

Seeing and Visualizing: It's Not What You Think, Zenon W. Pylyshyn, 2003

Evolution and Learning: The Baldwin Effect Reconsidered, Bruce H. Weber and David J. Depew, editors, 2003

The New Phrenology: The Limits of Localizing Cognitive Processes in the Brain, William R. Uttal, 2001

Cycles of Contingency: Developmental Systems and Evolution, Susan Oyama, Paul E. Griffiths, and Russell D. Gray, editors, 2001

Coherence in Thought and Action, Paul Thagard, 2000

From Signal to Symbol

The Evolution of Language

Ronald J. Planer and Kim Sterelny

The MIT Press
Cambridge, Massachusetts
London, England

The MIT Press would like to thank the anonymous peer reviewers who provided comments on drafts of this book. The generous work of academic experts is essential for establishing the authority and quality of our publications. We acknowledge with gratitude the contributions of these otherwise uncredited readers.

This book was set in Stone Serif and Stone Sans by Westchester Publishing Services. Printed and bound in the United States of America.

Library of Congress Cataloging-in-Publication Data

Names: Planer, Ronald J., author. | Sterelny, Kim, author.
Title: From signal to symbol : the evolution of language / Ronald J. Planer and Kim Sterelny.
Description: Cambridge : The MIT Press, 2021. | Series: Life and mind: philosophical issues in biology and psychology | Includes bibliographical references and index.
Identifiers: LCCN 2020041673 | ISBN 9780262045971 (hardcover)
Subjects: LCSH: Language and languages--Origin. | Language acquisition. | Human evolution.
Classification: LCC P116 .P585 2021 | DDC 401--dc23
LC record available at https://lccn.loc.gov/2020041673 ISBN: 978-0-262-04597-1

10 9 8 7 6 5 4 3 2 1

For Eric, my son
—R.P.

For my daughter Kate with love, and with appreciation of her tolerance of her father's eccentricities
—K.S.

Contents

Preface

This book has had a long gestation. One of us (Sterelny) has been working for the best part of fifteen years primarily on the evolution of human social life and on the capacities that make that life possible. As that work progressed, he became increasingly oppressed by the realization that a fully serious treatment of language was inescapable. He devoted a chapter of *The Evolved Apprentice* to the evolution of human communication in general, arguing there that the specific issue of deception as a threat to stable communication has been oversold, once one had an account of the stability of human cooperation in general, and once one takes fully into account the difficulty of project-managing deception in many-many interactions with much communication being about the here and now. However, with the exception of a defense of a gesture-first picture of the emergence of language, neither that chapter nor associated work burrowed deep into the specifics of language and its proximate ancestors.

So while Sterelny continued to build a detailed conception of human social evolution, language-specific issues remained somewhat on the back burner until Planer arrived in 2015 and we began to collaborate, with our picture of the book itself taking shape in 2018. Writing the book—the actual generation of text—has been thoroughly collaborative and approximately equal. We have both worked on revising the whole text, with Sterelny having written the first draft of the initial chapters and Planer of the later chapters. Intellectually, it is fair to say that the general conception of human evolution that structures the overall argument derives more from Sterelny, whereas the language-specific elements, and especially the ideas relating to cognitive neuroscience and the evolution of structure, owe more to Planer. Given that, he should be regarded as first author of the monograph. We will now turn to a brief picture of the book as a whole.

Language is foundational to human cognitive and social life. Hence, no adequate evolutionary account of the uniqueness of *sapiens* can escape the challenge of language evolution. As many have noted, this challenge is exceedingly difficult, in part because the gap between animal signaling and human linguistic communication is so glaringly large, and in part because evidence of the antecedents of language is elusive and indirect. Nevertheless, a growing number of researchers have begun to tackle this formidable challenge (whom we join here), and we think with valuable results. So we will begin by locating our own ideas against this cohort.

In the last decade or so, in addition to a new journal and various edited collections, we have seen important and ambitious books on the evolution of language. These include: Hurford (2007), Tomasello (2008), Bickerton (2014), Hurford (2011), Progovac (2015), Scott-Phillips (2015), Berwick and Chomsky (2016), Everett (2017). One major divide among these researchers is that between those defending a thoroughly gradualist model of language evolution and those who think language appeared much more suddenly (e.g., Berwick and Chomsky 2016). We figure firmly in the first camp, but that is still a diverse group. In our view, any adequate theory of language evolution must satisfy two important constraints. It must identify a plausible evolutionary trajectory from great-apelike communicative abilities to those of modern humans, where each step along the way is small, cumulative, and adaptive (or at least not maladaptive: there might be some role for drift). In other words, it must offer a "lineage explanation" (Calcott 2008) for language. Second, that account must be embedded within a larger account of human social, technological, and economic evolution, one enjoying independent evidential support from paleobiology, archaeology, and evolutionary biology. We have quite rich evidence of how and where ancient humans lived and moved. Information about their lifeways—about the resources they consumed, the environments they could exploit, the risks they could ameliorate, the territories over which they moved—constrains accounts of their cognitive, social, and communicative capacities (Sterelny 2016a). It is difficult to construct a coherent account that is both incremental and embedded, and any theory satisfying these requirements is no mere "just so" story.

Another major dividing line concerns the role that theory of mind played in the evolution of language. For some researchers (e.g., Tomasello 2008; Scott-Phillips 2015), the evolution of language is largely explained by a grade shift in theory-of-mind capacities. More specifically, on this view, the creation of language followed more or less inexorably once our ancestors were able to act with and recognize (notoriously complex) Gricean communicative intentions. The opposite extreme is also to be found in the literature, with some researchers understanding uniquely human theory-of-mind capacities to be the product rather than the cause of language. These researchers tend to be more concerned with the unique structural properties of language—in particular, its syntax and the basis of syntax in the human brain. As will become clear in the chapters that follow, here we very much occupy a middle position. In short: yes, an increase in theory of mind was important, but not necessarily for the reasons that those in the first camp typically cite. Moreover, an increase in those capacities was not the only critical intrinsic cognitive change that made language as we now know it possible. One way in which our view differs from others is that in general we are skeptical of single-factor, crucial-adaptation accounts of the evolution of language, just as we are of the emergence of the human mind and human social life more generally. Human language, and the cognitive and social capacities that support it, differ from other animal communication systems, including those of great apes, in a number of important ways. These include: the open-ended structural complexity of utterances (and here syntax is indeed critical); its diachronic flexibility (any competent user can coin new words and expressions); the great variety of social and communicative functions it can serve; its expressive richness, and the contextual sensitivity of that richness; the extraordinary speed, reliability, and agility of multiperson conversational interaction; its power as a tool of social learning and organizer of thought, for the division of linguistic labor makes it possible for us to think things because we can say them. We doubt that any of these features of language, with their supporting features of mind, are privileged: the single crucial difference from which all else follows. Instead, we think the evolution of language began from a mosaic of initially smaller differences from baseline capacities, and their interaction drove the evolution of communicative systems that became increasingly unlike any seen elsewhere in animal life.

In much of the work on human life, biology is contrasted with culture, with supposedly biological features of our lives and bodies contrasted with cultural ones. Language is an ideal case for undermining that pernicious dichotomy: language is both pervasively biological (supported by adaptations based on genetic changes, though often ones not specific to language) and also shaped by cultural processes on time scales from tens of thousands of years to minutes, as a witty new term catches on. Over the long duration of human deep time, there has almost certainly been a directional shift in which more of human communicative activity, and especially human vocalization, has been bought under top-down control: words are not much like snorts or grunts. But this should not be seen as a transition from biology to culture. Our eating is under top-down control, allowing what, when, and with whom we eat to be profoundly influenced by cultural learning. How, what, and with whom we eat is of great social significance. But that we have to eat, and what we can eat, are matters of morphology and physiology, though in this case a morphology and physiology that have been shaped by the cultural invention in the deep past of cooking. Eating is thus strongly biocultural. Laughter is another clear example of a biocultural trait, though it interestingly displays limited top-down control: genuine laughter is involuntary, and yet the triggers for laughter are profoundly influenced by culture. Language is typical rather than atypical in having a biocultural character.

One of the lessons of evolutionary and ecological theory is that life is messy, and this includes human life and deep human history. Biological kinds are fuzzy, without sharp boundaries; biological processes are typically driven by multiple interacting causal factors, and changes in those processes are often changes in relative importance. While formal models rightly idealize away much of this complexity, their results need to be interpreted in ways that reflect it. Our narrative is built around this recognition of multiple factors, with change in their relative importance over time. Our approach differs from others in a second way, one that reflects our disciplinary origins. Our argument is grounded in deep history rather than linguistics. Many of the books on language evolution to date have a shared approach that reflects the fact that they have mostly been written by linguists. These works tend to begin with a favored conception of language as it now is and work backward, attempting to identify the origins of language through the lens of their theoretical framework. Linguistic data are their empirical bedrock. For example, Progovac's monograph embraces a specific theory of the syntax of human language in general, and develops an account of how such a syntax could be assembled incrementally. Since we do not commit to any specific account of the character of natural language syntax, our account could in principle complement hers. For we bring an alternative strategy: we attempt to identify the lifeways of ancient hominins in enough detail to sketch the cognitive and communicative capacities on which those lifeways depended. These lifeways changed over time, and in favorable cases we have sufficient information to outline the capacities and needs implied by these changed lifeways. So, we aim to sketch a plausible and empirically constrained sequence of changing communicative lives, beginning with lifeways not much more collaborative or complex than living great apes, and ending with lives whose social complexity and expressive needs approximate those of living humans. To the best of our knowledge, this is the only monograph-length model of the evolution of language organized around and embedded in a specific, reasonably detailed, empirically constrained view of changing hominin lifeways. Indeed, perhaps the most important difference between the approach exemplified in this book and those mentioned above is the extent to which our conception of the evolution of language is embedded within an archaeological context, and a more general account of the evolution of human sociality. Everett, for example, defends a view of the incremental emergence of language which is in some ways broadly similar to ours, but which places the establishment of something close to full language much earlier in time. We see no independent archaeological support for the existence of the very impressive cognitive capacities human language requires a million or more years ago. Likewise, Bickerton's final foray into this problem was an attempt to solve "Wallace's problem": the supposed fact that humans are smarter than they need to be (and this includes our language abilities) to cope with the problems posed by our physical and biological environment. Wallace overlooked the fact that we have to be smart enough to cope with one another, and that this can generate positive feedback, ratcheting up our cognitive capacities. Bickerton, in our view, developed a solution in search of a problem.

We said above that we are not in the hunt for a crucial breakthrough, a new adaptation that took our ancestors into a new adaptive zone. In our view, rudimentary forms of all or most of the elements on which language depends are found in the great apes and were part of the equipment of the earliest species in our lineage. The elements include: theory of mind (as mentioned above); executive control, providing the ability to plan and execute complex and precise action sequences; the ability to notice and parse action sequences performed by others; but also prosocial motivations of various kinds, since conversation is typically a form of cooperation. This view of language evolution gives a prominent role to intermediate language forms, such as early protolanguage, enriched protolanguage, and early language, and to the flexibility and adaptability of hominin minds, even ancient hominin minds. By recognizing these intermediate language forms and the intermediate minds supporting them, we can begin to close the gap between great ape communication systems and language without positing evolutionary miracles along the way; we can, in other words, begin to glimpse a lineage explanation for language.

But as we also said, any adequate account of the evolution of language must be embedded within a broader frame of hominin evolution. We argue that the broader context in which language evolution played out is usefully conceived of in terms of two cooperation revolutions. The first of these connects great-apelike social behavior with that of early *Homo*. It marks the transition from the rather limited forms of cooperation characteristic of chimpanzees to a regime of obligate mutualistic cooperation (Tomasello, Melis, et al. 2012; Tomasello 2014). This period saw an initial upgrade in several cognitive abilities critical to more sophisticated forms of communication—enhanced social learning, enhanced theory of mind, enhanced memory, future-directed action, and more. The second revolution connects the already quite complex lifeways of late *erectus* and *Homo heidelbergensis* with that of modern humans. This revolution involved the appearance of regular, delayed-return forms of cooperation, such as giving a person a tool today in exchange for some meat next week (Sterelny 2014). Life is even more complex if your generosity with your tools is rewarded by help from a third party, recognizing your reputation as a reliable and helpful member of your community. We argue that the shift to delayed-return cooperation introduced a range of new social challenges. It is at this stage that tracking the reputation of third parties via gossip or its equivalent became essential to the stability of cooperation, and this drove the need for more complex communicative technologies. If gossip matters, and if it matters that gossip is accurate, those who exchange gossip need to be able to specify what happens off stage—who did (or failed to do) what to whom—precisely and unambiguously.

This shift to a more efficient but more complex and challenging form of cooperation within hominin bands or residential groups probably partly overlapped in time with an expansion of the social and spatial scale of cooperation, as residential groups became more open to one another, with freer movement in and out, and as they became networked into larger communities.

This too led to important benefits (information exchange, a larger reproductive market, demographic buffering). But once again it made the social and communicative landscape more complex (Sterelny 2021).

In telling our particular story, we will draw upon some familiar ideas. Among them: (i) that gesture played a key role in launching human language evolution (Donald 2001; Tomasello 2008; Corballis 2011; Sterelny 2012b); (ii) that stone toolmaking drove the evolution of human syntactic abilities, and plausibly other cognitive abilities linked to language (Stout 2011; Stout and Chaminade 2012; Planer 2017b); (iii) that large-game hunting and the control of fire drove increases in communicative complexity by intensifying demands on human cooperation and coordination (Pickering 2013); (iv) that the evolution of singing in our line played an important role in preadapting humans for vocal language (Gamble, Gowlett, and Dunbar 2014; Killin 2017a; Killin 2017b), with the increased control of fire providing an ecological context selecting for the transition to the vocal channel; (v) that the human "release from proximity" was facilitated by the evolution of complex kin terminologies and other linguistically enabled ways of keeping track of relationships between people (Gamble 2013; Planer 2020a). In developing our conception of the evolution of language, we bring these ideas together into a novel, coherent package, while also updating and extending them in ways supported by more recent paleontological, archaeological, phylogenetic, and genetic evidence. Or so we hope!

With this general picture sketched, we now present a brief summary of the book's chapters. In chapter 1, we develop our view of the constraints that a theory of the evolution of language should satisfy, building on the work of Laland (2017), and then commit to an incremental view of the evolution of language in which language evolved through a series of increasingly rich protolanguages. We then discuss skeptically one popular incremental conception of language evolution, the idea of progress through indexes and icons to true symbols. Models of the evolution of language have often appealed to the index-icon-symbol distinctions, suggesting that the puzzle of language is tied to the problem of explaining the emergence of the cognitive capacities and social environments that support symbolic communication (e.g., Deacon 1997). We argue that the index-icon-symbol distinctions are much less clean and transparent than has been supposed. We prefer the sender-receiver framework, much discussed in the recent literature, as our organizing conception of communication and its evolution. This setup chapter ends with our proposal about empirically constraining pictures of language evolution.

Chapter 2 begins the substantive project. We begin by using the resources of comparative biology and paleoanthropology to identify the cognitive and communicative baseline: the capacities of early hominins. That is one end of the pathway; in the second section, we give our characterization of the language mosaic, the cognitive and social abilities that support language. This is the other end of the pathway. The critical claim about this mosaic is that few, if any, of these capacities are unique to humans or to language. That matters. For if that claim is right, most, perhaps even all, are much-elaborated versions of capacities that existed in somewhat rudimentary forms in earlier hominins. Moreover, since these capacities feed into many skills, there can be a historical signal of their existence, for they can be manifest in behaviors like foraging and toolmaking, behaviors that do leave physical traces. Having identified the two bookends, this chapter elucidates the first advances of Pliocene hominins beyond great ape capacities, with the core idea being that a shift to obligate bipedal lifeways brought with it important cognitive and behavioral changes, including, in particular, changes to cooperation, coordination, and communication; in our view, via an expanded role for gesture.

Chapter 3 further articulates our view of the early stages of language evolution. This chapter takes as its principal question the transition from animal signals to words. Animal signals tend to be triggered by specific environmental states and linked to specific behavioral responses. They are in general tied to the here and now and tend to be produced unaccompanied by other signals. Moreover, the stock of signals belonging to an animal communication system tends to be fixed over time, whereas human lexicons are readily expandable. We explore how each of these constraints might have been lifted, yielding words or at least wordlike elements. These changes take place, in our view, against the background and in the aftermath of the first cooperation evolution. We estimate that some form of protolanguage using wordlike elements was very likely ancient, dating at least back to cooperative hunting, which is probably at least 1.7 million years old. As the hominin ecological repertoire expanded very gradually over time, we claim that the lexical stock expanded with it. So our crucial claim in this chapter is that erectine hominins, the hominins that appeared after about the first third of the Pleistocene, were probably equipped with quite rich protolanguages. These were protolanguages with structured signs, with displaced reference, and with enough flexibility to add new items over time. We aim to show that our information about those lifeways supports the view that these hominins were both capable of using a protolanguage and very likely needed a protolanguage, one with flexibility and displaced reference. We introduce the link between gesture and structured signs in chapter 3, but structure is the focus of chapters 4 and 5.

More specifically, chapter 4 addresses the origins and establishment of what we call "composite signs"—signs with less than basic syntax but which are an important precursor to basic syntax. Composite signs have parts that work together in some way to fix the meaning of the whole, but the order of the parts does not yet encode meaning. In developing this view, we respond to an important challenge raised by Thom Scott-Phillips (2015). He argues that the introduction and use of composite signs requires very rich cognitive capacities. We argue that a somewhat simpler cognitive skill set suffices, a view that supports a more incremental model of the emergence of structured signs.

Chapter 5 takes up the evolution of syntax proper. In particular, it addresses the evolution of hierarchical structure. Despite the attention hierarchical structure has received in the evolution-of-language literature (see, e.g., Stout 2011; Berwick and Chomsky 2016), exactly what such structure amounts to is far from clear. So one goal of this chapter is to offer an improved analysis of hierarchical structure and its role in communication. We then develop the idea that the computational machinery underpinning hierarchical structure evolved in the service of technological skill, the production of sophisticated stone tools in particular. (As our language here suggests, in this chapter, and throughout the book more generally, we commit to a broadly computational picture of cognition. We assume rather than defend this orientation in the book.) Toward this end, we consider evidence at the cutting edge of both lithic studies (pun intended!) and the neuroscience of complex, intentional action, and attempt to integrate the two. We suggest that, by the time composite tool production had become part of the human technical repertoire (between 500 and 250 kya), most if not all of the cognitive specializations for processing hierarchical structure were in place. So the central strategy of this chapter is to show that the cognitive capacities needed to master the impressive stone technologies of the erectine and especially heidelbergensian hominins (appearing at about 800 kya) could be, and probably were, co-opted for structured language processing. These tools could only be made through an accurate sequence of planned steps, and this skill could only be learned from others by being able to observe and recognize such sequences of planned steps. The advanced stone-working techniques of these hominins depended on hierarchically represented action plans, both in their execution and acquisition. In our view, likewise, this is the cognitive capacity on which producing and understanding hierarchically structured sentences ultimately depends.

Chapter 6 takes up a crucial challenge for any view of language that begins with an expanded role for gesture. If protolanguages began as largely gestural systems, why and how did vocalization become so important? We meet that challenge through the idea of a firelight niche, and the changed social and physical environments that come with the control of fire. In developing this view, we adapt some ideas of Dunbar in identifying the selective forces that favored increased vocal control. In our view, selection for something like wordless singing and likely laughter led to improved vocal control. These behaviors helped to ease tensions and strengthen affiliative bonds as hominin social life became more complex and intense. With more vocal control available, the vocal channel offered various efficiencies, and we argue that those were particularly salient at the fireside, in the firelight niche. So we think a combination of increased vocal control (evolving independently of protolanguage) and the control of fire mediated the shift to a primarily vocal mode of communication.

Chapter 7 rounds out our story. It is organized around changes in human lifeways late in the Pleistocene, with the second cooperation revolution in full swing, perhaps beginning something like 150 kya (this claim is defended in detail in Sterelny 2014; Sterelny 2021). This second revolution both increased the demands on human communicative capacities and provided an environment in which cultural learning was more efficient. Jointly, these changes explain the transition from richer protolanguages to languages as we know them (though in our view there is no sharp boundary between the two). So this chapter presents a detailed account of the social tools and communicative capacities on which this second revolution depended. These include kinship terminologies, explicit norms, and the capacity to tell stories and report indirect speech. In our view, it is not until around this period that human lifeways began to closely resemble those of ethnographically known foragers in terms of the cooperation and coordination problems routinely faced and solved. In turn, we claim that it was probably not until around this period that fully modern language began to appear, evolving to meet the conflict flashpoints associated with intensified cooperation and coordination demands on humans.

While some might see this as a strikingly late date for the evolution of language, we see this as a natural consequence of taking the cultural-evolutionary character of language seriously. The genetic prerequisites for language are probably much more ancient, given the rather minor genetic differences between *sapiens*, Neanderthals, Denisovans, and hence, by inference, their immediate ancestor, the heidelbergensians (Dediu and Levinson 2013; Levinson and Dediu 2018). But we think the role of cultural evolution was of critical importance. Compare language with numerical cognition. It is one thing to have the genetic capacity to multiply 127 by 69; another to have that as part of one's cognitive phenotype. In the middle Pleistocene, and in the lives of the heidelbergensians, there was little need for the full package of design features characteristic of modern language. Moreover its evolution may well have been prevented by demographic constraints. Our ancestors of 500 kya or even earlier may well have been "language-ready"—indeed, we think they probably were—but language readiness does not mean they were language-equipped.

Chapter 8 is self-assessment: we return to the success conditions identified in section 1.1, and evaluate our picture in the light of those criteria. While we certainly do not fully meet them, we claim to have sketched important parts of the map by offering empirically constrained and incremental models of an expandable lexicon, displaced reference, and the cognitive prerequisites of hierarchical syntax.

We now turn to a pleasurable duty: acknowledging and thanking help along the way. It takes a village to raise a book, and we have been fortunate in our village. The local Australian National University (ANU) environment has been very supportive. That is true both of the School of Philosophy and of the interdisciplinary (though linguistics-heavy) Centre for the Dynamics of Language. In addition to housing, supporting, and tolerating us, these have been places where both authors tried out, individually and jointly, presentations that were early versions of these chapters (especially at assorted events organized through the Centre), and we have had a lot of helpful feedback: in particular from Carl Brusse, Nick Evans, Liz Irvine, David Kalkman, Anton Killin, Stephen Mann, Richard Moore, Ross Pain, Lauren

Reed, and Matt Spike. These robust individuals read and responded to draft papers and suffered through presentations. We have also tried out various chunks at workshops in Australasia (in particular, at an evolution-of-language workshop organized at ANU; assorted CoEDLFests, the Wellington Empirical Philosophy workshops, and a series of archaeology-meets-philosophy workshops). For help at these, we would particularly like to thank Peter Godfrey-Smith and Peter Hiscock. They too responded to both presentations and drafts, and they have both contributed in essential ways to how we think about evolution, language, and theory-building in the historical sciences. In addition, four readers for the MIT Press read the whole manuscript, and their comments provoked many improvements. Two were nameless. The other two were Michael O'Brien and Richard Moore. We thank all four, but Richard Moore in particular provided an especially searching set of comments. Finally, we thank the Australian Research Grants Council for their support of this project in particular, and for their support of Sterelny's projects over many years.

1 An Intractable Challenge?

1.1 The Scope of the Problem

Human language confronts the community of evolutionary theorists with a difficult but inescapable challenge. Inescapable, since language is manifestly central to human life, and so an account of the origins of human cognition and social life must include an account of the emergence of language. Moreover, that account must be evolutionary. That claim is not controversial, though there is debate about the extent to which the emergence of language was gradual and incremental (rather than abrupt), and about the relative roles of genetic and cultural evolution. For even if language was an invention, as writing clearly was, it was invented because of evolved features of hominin life, and invented using evolved features of hominin cognition. Moreover, even if language was an invention, almost certainly it was invented incrementally and cumulatively, exemplifying cultural if not genetic evolution (Heyes 2018). We will return to those debates shortly.

The challenge is difficult, because language is very different from other forms of communication. No other extant species uses a minimal or rudimentary version of language, and so it is difficult to exploit the resources of comparative biology to identify rudimentary forms of language, or the environmental features that select for its evolution. Difficult, also, because the explanatory target is in part a feature of the social environment, in part a feature of individual psychology. Evolutionary biologists distinguish between an organism's genotype and its phenotype, the array of morphological, physiological, and behavioral traits influenced by its genes. Humans' extraordinary large and complex brain is one such phenotypic trait; speaking English is not. Specific languages are created, maintained, and modified over time by the communities that use them. The authors of this book are both native English speakers, and in an important sense, in our respective childhoods, we *joined* English. We did not create it *de novo* from inherited resources; say, inherited instructions about how to build a language. In contrast, ordinary phenotypic traits are created *de novo*. We do not inherit mini-legs, or grab a pair of readymade legs from the environment. Our physical organs were built *de novo* as we developed from zygote to toddler. In contrast, our language was a persisting feature of our social worlds to which we adapted. But the capacity to speak and understand English *is* an individual trait, a feature of each of our (the authors') realized phenotypes. Evolutionary theories of language sometimes take the social phenomenon as their primary target, sometimes the individual trait. An adequate theory needs to give an account of both, and of their interaction. A further difficulty is an ongoing and deep controversy about the nature of the explanatory target itself. There is much debate about the evolution of our large and complex brain. But at least there is consensus on what the brain is. Evolutionary theorists of language have no such luxury. The Chomskian idea that language is primarily an organ of thought, only secondarily exapted for communication (Berwick and Chomsky 2016), is just one example of a controversial claim about language of obvious relevance to theories of its evolution. In general, an account of how language evolved will typically be shaped by an account of what language is, and there is no consensus about that. Finally, and most obvious of all, our evidential basis is largely indirect. We have to infer the scope and limits of ancient hominins' communicative and cognitive capacities from the traces left by the actions those capacities made possible. For having (or lacking) a language leaves no clear signal on the skeleton of a language user (and hence on the potential fossils of such users). It is true that speech depends on extraordinarily complex control of breath, tongue position, and mouth shape (Fitch 2010; Everett 2017, chapter 8), and there are fossil indicators of such control, as we shall see in chapter 6. But the capacity to produce complex streams of sound might reflect the ability to sing rather than speak, and language can be manifest in sign rather than sound. Evidence of breath control is important, but most evidence about ancient language is indirect.

Despite some skepticism, we think the case for an incremental conception of the evolution of language is overwhelming. For one thing, the cognitive capacities required for routine conversation are extraordinarily impressive. The speech stream typically flows at somewhere between 135 and 180

Box 1.1

A Note on Terminology

"Hominin" refers to all members of the lineage to which we belong, and which began when the last common ancestor of *H. sapiens* and the chimp species split. The chimp species are the only surviving representatives of one lineage, and our species is the only surviving representative of the other lineage, collectively known as the hominins. As do many others, we use "human" as an informal term for the recent, exceptionally large-brained hominins. We use "AMH" as an abbreviation for anatomically modern humans, humans morphologically grouped with our species, first appearing in the fossil record about 300 kya (kya = thousands of years ago; similarly, mya = million years ago).

words per minute (Everett 2017, chapter 8). In managing that stream, the speaker has to decide what he or she wants to say; compose that thought into a specific ordered and structured set of lexical items; and turn that representation of what to say into a sequence of motor commands that drive articulation. This cognitive feat is executed under time pressure, as participating in a conversation requires one to match its normal flow seamlessly, and the utterance the speaker launches must be both responsive to and constrained by the previous contributions to the conversation. Inserting a comment requires the speaker to be tuned, at least to some extent, to the attention, interests, and knowledge of the audience.[1] In the evolution-of-language literature, there has been a recent focus on pragmatics, making much of the fact that what a speaker intends to convey (and usually does convey) is not well captured by the literal meaning of the utterance.[2] Ron says "Are there any further questions?" and his audience immediately realizes that he is suggesting that the group adjourn to the pub. Ron's choice of lexical items is shaped on the fly by his representation of how his audience will hear and interpret his words.

In part for this reason, listening and interpreting is at least as demanding as speaking. In listening, the agent hears a continuous and idiosyncratic speech stream (as every voice is different) as a discrete series of words, a skill developed and automatized early in language acquisition. In listening to a conversation in an unknown language, it is impossible for most of us to identify the words or phrases within the stream. Armed with this skill, the listener must identify the lexical choices of the speaker, including the organization of the words into phrases and clauses. The listener must do so in the face of the rapid flow of speech and its extremely ephemeral nature, for human sensory memory decays very quickly. As noted above, it is not sufficient to identify the lexical items in their structure: the listener needs to identify what was intended, not just what was said. Moreover, Stephen Levinson has shown that interpretation depends on prediction (Levinson 2016). The gap between one utterance finishing and the response beginning is so brief that the response must be at least partially formulated for launch before speakers finish their contribution. Listening involves predicting what the speaker is saying and when he or she will finish saying it. Formulating and beginning a response (with all its challenges) takes place partially in parallel to listening and interpreting.

Conversation thus depends on a complex of cognitive capacities, some of which are very demanding. Nothing in great ape communication looks remotely as challenging. It would be extremely surprising if these capacities arrived in a single evolutionary step. Some of those working on the evolution of language draw a distinction between a "faculty of language" broadly conceived and one narrowly conceived (Hauser, Chomsky, et al. 2002). The idea is to distinguish between cognitive capacities that are relevant to language, but which play other roles in the life of the agent, and those capacities that are distinctive of language. So (the thought runs) theory-of-mind capacities, a powerful and retentive memory, executive control, counterfactual reasoning, and the like are all essential to the use of language as a means of communication in a community, yet they play many other roles in human life. These capacities are probably not uniquely human, but they have been upgraded in the course of hominin evolution. An incremental view of their evolution is not seriously challenged. In contrast, consider the mastery of syntax: the ability to compose lexical items not just into a sequence but as a hierarchically organized structure with no upper bound in complexity. This ability is not just essential to language but may be unique to it. Perhaps the same is true of the ability to use and understand terms that have displaced reference (Deacon 1997). So while the multifunctional tools in the faculty (theory of mind and the like) have incremental evolutionary histories, perhaps (the thought goes) the language-unique capacities (in particular, syntax) evolved in a single step.

We are skeptical of the distinction between broad and narrow conceptions of the faculty of language, in part because it understates the central role of the multiuse capacities for linguistic competence, and in part because we will suggest ways of connecting syntactic composition and displaced reference with other aspects of human cognitive and social life. Moreover, for reasons that we will explain later, we are very skeptical of the idea that the mastery of syntax could emerge in a single step. However, even if we are wrong about this, we still need an incremental theory of the evolution of almost everything else that is essential to the human mastery of language. We will return to this point in a little more detail below.

Given all this, what should we aim to do? Kevin Laland has suggested that any credible theory of the evolution of language should satisfy the following criteria (Laland 2017):

The theory should explain the honesty of early language;

The theory should depend only on well-understood evolutionary mechanisms;

The theory should explain the distinctive scope and expressive power of human language;

The theory should account for the uniqueness of human language; and

The selective pressures proposed by the theory that led to language should be consistent with the known variability and dynamism of hominin environments; there is no one environment, physical, social, or biological, in which hominins evolved. Indeed, hominin evolution resulted in the transformation of hominin social, biological, and physical environments, and hence of the selective pressures to which our ancestors were exposed.

We agree. But in constructing a thoroughly gradualist model, an additional criterion needs to be made explicit, a criterion nicely expressed by Brett Calcott's idea of a "lineage explanation" (Calcott 2008). Such an explanation identifies a sequence of changes that begin with great ape communicative capacities (which we take to approximate the capacities of the last common ancestor of the *Pan* and hominin lineages), ending with contemporary human capacities. A candidate lineage explanation must meet two further conditions. First, each stage in the sequence must vary from its predecessor and successor only in relatively minor ways: the links in the evolutionary chain that connects great-apelike capacities to ours must each be one plausible evolutionary step apart. Second, no step can be blocked by selection. In principle, a successor step could drift to fixation, but the normal expectation is that a successor state confers a fitness advantage on its bearers in the environment of the time.

Box 1.2

A Note on Geological Epochs

Name	Approximate Dates	Remarks
Miocene	23.03–5.33 mya	The first hominin fossils are found late in the Miocene, but they are so fragmented and scattered that very little can be firmly inferred about the lifeways of these hominins.
Pliocene	5.33–2.58 mya	By the last 500k years of the Pliocene, and likely much earlier, there were bipedal hominins using stone tools. Our analysis of hominin lifeways, and their implications for hominin communication, begins with these hominins.
Pleistocene	2.58 mya–12 kya	This epoch was characterized by significant and increasing climatic instability, with accentuating glacial and interglacial cycles. It was the epoch in which there was massive encephalization (relative growth in size of the brain and the neocortex) in the hominin lineage, and great and cumulative changes in hominin lifeways. Most of our analysis is focused on Pleistocene hominins.
Holocene	12 kya–present	By the beginning of the Holocene, *sapiens* was the only extant hominin species. In our view, and in everyone else's view, language as we now know it has been a universal possession of every Holocene human community.

In the case of language evolution, the natural thought is that selection favored a gradual expansion of communicative abilities, but as we have noted, we need to bear in mind the possibility that some changes may have been selected for effects having little to do with communication. For many of the abilities on which language rests support other skills. The evolution of language is an example of mosaic evolution; a set of originally largely independent capacities coevolved, gradually becoming more integrated, but with many of those capacities continuing to play a role in our cognitive and social lives independently of language. The evolution of language is in part a history of upgrading existing capacities; probably in part a history of building new capacities; in part a history of integrating those capacities so they work smoothly together, often with great rapidity. As there is controversy about the exact cognitive demands of language, so too there is some controversy about the membership of this coevolving mosaic of capacities linked together in the emergence of language. We give our view of that mosaic in section 2.2.

At this point, it will help to give a map for the rest of the chapter, and, very briefly, of the rest of the book. This chapter is devoted to articulating and defending three framework ideas that will guide the discussion in the chapters to follow. The first of these is a commitment to a specific version of an incremental model of language evolution: the idea that language evolved via a sequence of increasingly rich protolanguages. Protolanguages are communicative tools that are language-like in having a lexicon, for utterances are strings of wordlike items. But they have little or no overt morphology or syntax. The second is a rejection of the typical incremental models of language evolution, ones that see language as evolving via a progression from indexes to icons (or from icons to indexes) to true symbols. In sections 1.3 and 1.4 we explain our skepticism, and then set out our preferred sender-receiver framework. In 1.5 we take up the vexed problem of empirical constraints on a theory of language evolution, and outline our partial solution to it.

As advertised in our preface, chapter 2 sets up our picture of the baseline capacities from which language evolved in our lineage, gives our view of the initial, substantial, shift away from great ape lives, which we take to have been in the Pliocene, and outlines our picture of the cognitive and social machinery on which language depends. Chapter 3 presents our case that the erectines evolved a gestural protolanguage. Chapters 4 and 5 take up the challenges of structured signals. Chapter 6 develops our account of the gesture-to-speech transition, and the role of the fireside in that transition. Chapter 7 delivers our view of the emergence of full language in the economically and socially complex world of the late Pleistocene. And finally, chapter 8 is review and self-assessment.

1.2 Protolanguage

Contemporary languages have a number of very distinctive features. One is that utterances are not just strings of ordered words. Words can be grouped into phrases, and those in turn can be grouped into still more complex structures within a sentence: there is hierarchical, not just linear order in sentence organization. Moreover, the recipe that organizes words into phrases can be applied iteratively, to produce constituents of increasing complexity playing the same overall role within the sentence. It follows that there is no clear upper bound to the complexity of a sentence in a language, nor to the number of distinct sentences that can be produced within that language. A second distinctive feature is that the lexicon is indefinitely and easily expandable: new words can be coined and added to the language at will, though languages impose various constraints on the sound and syllable sequences of possible words in that language. Third, there are semantic properties common to all languages. Displaced reference is always possible: in every language, speakers can talk about the elsewhere and the elsewhen, and hearers can understand those utterances.[3] In every language, a variety of speech acts are possible: stories can be told, jokes made, questions asked, rituals performed, information imparted. Fourth, all living languages have a form of dual coding. There is a base set of units—phonemes—that are used repeatedly and that can be reidentified, though their exact acoustic properties vary from speaker to speaker and context to context. These elements are not independently meaningful. But they are used to build a second set of items that are also used repeatedly and can be reidentified, but which are independently meaningful. These are the morphemes and words of a language. We think there is no reason to assume that these four characteristics appeared at the same time in the emergence of language. Moreover, at least three are gradient phenomena: a lexicon can be more or less readily expandable; dual coding can be partial or complete; the menu of distinct speech acts can be longer or shorter, and hence the system can have more or less expressive power.[4]

In addition to defending an incremental model of the evolution of language, at the beginning of our argument we will endorse one more substantive claim about its evolution. Following (especially) Derek Bickerton and Ray Jackendoff, we think that an important intermediate stage in language evolution was the establishment of protolanguage, or more exactly, a sequence of protolanguages (Lieberman 1998; Jackendoff 1999; Bickerton 2002; Bickerton 2009; Bickerton 2014). Protolanguage is a communication system in which agents produce and understand strings of wordlike terms: terms with displaced reference and other semantic features of words. The repertoire of such terms can be quite large. But the sequences are not syntactically organized, or are so only in a very rudimentary way. Our picture of protolanguage is pieced together from various sources. These include trading lingua francas that typically develop between maritime traders and their small-scale, linguistically diverse partners. Other sources are pidgins, systems that arise when people are thrown together over substantial periods and must communicate, but have no common language. These situations have arisen quite frequently in colonial systems of forced labor, and in situations of adult migration of groups to another linguistic community. Pidgin-like systems typically have quite extensive vocabularies but little or no grammatical or morphological structure, and their word order is often quite variable. Another source is emerging sign languages: village sign, community sign, and homesign (Meir, Sandler, et al. 2010). These sign systems differ in the social networks that use them, but they all involve the need of profoundly deaf individuals to communicate in the absence of an established sign language. In contrast to pidgins, however, the deaf individuals using these systems lack prior access to a conventional linguistic model. Village and community sign systems are quite dynamic (more on this in section 4.4), but because that dynamism is driven by agents with no access to a standard language, some have seen the early stages of these systems as a useful guide to early language evolution, especially those who think that the origins of language are gestural.

These are all face-to-face communication systems, typically somewhat restricted in their expressive power, with mutual understanding depending heavily on context. But care must be taken in using them to develop a picture of protolanguage. The creators of these systems may have built them with the aid of cognitive capacities that evolved only after (and because of) the inception of language. So, we assume that protolanguages were *no more* organized and expressively rich than village sign, pidgins, and the like. At least initially they were probably much less expressively rich and less organized.

On the picture we develop, then, language emerged via a series of protolanguages (with many existing at any one time), with those protolanguages—depending on their stage of development—differing dramatically in lexical richness and in their expressive power: for example in the range of speech acts they supported. We expect they also varied in the extent to which word sequences were regularized, with somewhat standardized ways of indicating agents and patients, the time at which events and actions took place (or will take place), and so on. There is no sharp line between protolanguage and language; for this reason, Everett, who otherwise endorses this aspect of an incremental theory, resists the term "protolanguage" (Everett 2017).

The referential elements in these prelanguage protolanguages were not quite lexical items as generative grammars represent lexical items. They were not (we presume) tagged with morphosyntactic features like number, gender, or transitivity. On the protolanguage-first view of the evolution of language, these morphology-and-syntax guiding features of lexical items (or the cognitive representation of these features of lexical items) came later.

Nonetheless, in their semantic properties these items were wordlike. They were reused, and when they were reused they picked out the same individual, action, or kind. In our view, something like words emerged early. This view is not universal. Steven Mithen argues that hominin communication until *sapiens* remained holistic, with a repertoire of calls that encoded information about a situation as a whole and/or instructions, but where the significance of the call as a whole did not derive from independently meaningful elements from which the calls were built (Mithen 2005; Mithen 2009). He and Allison Wray suggest that hominin communication systems remained holistic until late in the transition to language (Wray 1998; Wray 2002; Wray 2005).

It is true that animal signaling systems like the vervet warning calls are holistic in this sense. But their call repertoires are very limited. Even if we imagine a much-expanded repertoire, the system will remain very inflexible. There is no toolkit for building new messages as needed. For example, when a new agent migrates into the group, there will be no obvious way of communicating about what that particular individual has done or ought to do. Moreover, we think there is no plausible model of how a large system of holistic signals could gradually turn into a structured system. Mithen and Wray have suggested that structure, something like individual words, can emerge through a process of segmentation, as speakers notice an initially accidental similarity between a set of holistic utterances and the situations those utterances signify. They give a toy example, suggesting that it might turn out that the element 'ma' appears in a number of vocalizations, each of which maps onto a situation in which a woman receives resources. So holistic protolanguage speakers come to infer that 'ma' means something like 'female recipient' and a word emerges.

To the extent that this scenario is at all plausible, it depends on the example being cut down, idealized. If we really are imagining a large call repertoire, it would be more than a miracle if an array of recognizable units mapped consistently onto some specific feature of the context of use, just by accident. The more holistic calls in which something like 'ma' is embedded, the more likely it is that some of them have nothing to do with women.

Added to this, speakers would have to recognize these recurring units within the calls, and so the scenario seems to presuppose that the holistic calls have phonological or syllabic structure[5] even though there is no semantic structure. That would not be true if phonemes and words have coevolved, as part of the digitalization of the sound stream. As Dan Dennett points out, that is the most likely origin of something like phonemes and/or syllables. As the number of protowords expands, they will tend to crowd the space of short, easily distinguished sounds (or gestures). Digitalizing vocalization, composing protowords from reusable units that speakers and hearers can produce and recognize, helps manage that problem (Dennett 2017). Finally, the speakers must somehow not just recognize the acoustic units within the holistic calls, but recognize their regular co-occurrence with a feature of the environment onto which, purely by accident, the calls map. Indeed, an individual must not only notice this covariation; if she attempts to exploit it in communicating, she must expect others to notice it too. Putting these problems together, we doubt that there is any plausible model of how a large system of holistic signals can incrementally turn into a structured system. This skepticism in part motivates our account of the origin of words, or wordlike elements, and also of basic syntax. It is one of our main reasons for preferring a gesture-led account of the early expansion of communication, since a gestural system readily evolves structure, with specific gestures having independent meaning.

1.3 Icons, Indexes, and Symbols

We are far from alone in thinking that we need an incremental model of the evolution of language, and for that reason a good deal of work on the evolution of language is organized around a framework derived from C. S. Peirce (though usually discarding most of his theoretical baggage), dividing signals into indexes, icons, and symbols (for two prominent examples, see Deacon 1997; Everett 2017). This framework looks as if it fits naturally into an incremental, cumulative conception of the evolution of language, though we think that is something of an illusion. An index picks out its target by being a causal product of its target (or perhaps just through being reliably correlated with its target) and is closely associated with its target in space and time (the root concept of an index being that it points to its target). An icon is a signal that picks out its target through some natural resemblance between signal and target. Symbols are often defined negatively: they do not depend on resemblance or regular association in picking out their target. Sometimes it is said that the relation between a symbol and its target is arbitrary and depends on convention. While that is far from illuminating, the general thrust of this framework is that indexes or icons are in some way simpler and more fundamental than symbols, and that index- or icon-using communication systems evolved before symbol-using ones, and were a precondition of symbol-using systems appearing.

It is generally assumed that the use of symbols is more cognitively demanding than that of either indexes or icons (most emphatically in Deacon 1997),[6] but there is some disagreement about which of the latter is more difficult. For Peirce and his close adherents, icons are the least cognitively demanding of all. As a general rule, this seems wrong. Animal signals are usually most naturally seen as indexes: the dance of a male lyrebird advertises his health and vitality, and does so by being causally dependent on that vitality. Moreover, there are many iconic signals which human infants only come to grasp long after they demonstrate a mastery of pointing.[7] Finally, the archaeological record suggests that indexical signals appeared much earlier than iconic ones in human history. For example, Michael Rossano (2010) has argued that handaxes very likely served as indexes of desirable but hard-to-perceive qualities of their makers (e.g., good eyesight, good concentration).[8] As he points out, these items appear over a million and a half years earlier in the record than do any plausible icons. The earliest supposed icons are the Tan-Tan figurine and the Venus of Berekhat Ram: stones which, some argue, have been modified to accentuate their anthropomorphic appearance. Both of these are dated to after 400 kya, and in any case their interpretation is highly controversial. The oldest undeniable icons are the cave paintings of Pleistocene Europe, and these are younger than 40 kya. Of course, paintings are archaeologically fragile, and much older ones probably existed. But the use of ochre only dates back to about 300 kya, and even if those earliest uses were as a pigment (ochre has other uses), pigments have other uses. Bodily decorations, for example, need not be icons. Australian aboriginals use striking patterns of black and white pigment as bodily decorations in their rituals, but those patterns are not icons. In any case, even if we push icons back to the earliest uses of ochre, indexes are still probably much older.

Dan Everett (2017) is another theorist who, while endorsing a broadly Peircean framework, rejects the icon-first view. He thinks that as our ancestors became equipped with an increasingly humanlike mind, icons were regularly added to an existing stock of indexical signals, followed by symbols. Language as it now is rests almost exclusively on the mastery of symbols, with just (in spoken language) a vestigial amount of iconicity with onomatopoeic words. Sign languages too are largely symbolic, but with a larger residue of iconicity. This progressive picture depends on the fact that indexes are produced in direct confrontation with their targets: standard examples are the mating display of the superb lyrebird, or the famous danger-specific warning calls of vervets. The vervets' eagle call announces the threat of an eagle here, now. Supposing that selection has not already wired in the right responses, the significance of calls of this kind can be learned by association, especially as the relevant feature of the environment is so highly salient. Associative learning probably explains how one animal learns to take advantage of other species' warning calls. Notice, though, that the distinction between icons and indexes is not very sharp, and this is one reason for our reservations about this framework. A dog's baring its teeth and growling is an index and causal product of its readiness to attack; it is a warning of a threat here, now. But it is also iconic, since it also resembles a dog that is about to attack, and that resemblance may in part explain how the display was recruited as a warning signal.[9]

In contrast to indexes and symbols, icons pick out their targets in virtue of a similarity between the signal and its target: maps, diagrams, pictures are paradigmatic icons. So are mimes, as when a traveler, lacking the appropriate language, mimes drinking in order to indicate the target of a request. Similarity need not be visual: many foragers are excellent vocal mimics of their local fauna, and so can, for example, iconize the location and identity of a bird by pointing to its location and mimicking its call. Iconic signals need not be restricted to the here and now. Miming a spear-throwing action might well (for instance) be a suggestion to hunt tomorrow; likewise drawing a mud map of some part of the local terrain can coordinate movement through that terrain. But an icon can certainly be like an index in being about the here and now. Male chimps often solicit sex by displaying their erect penis to prospective partners, a signal about the immediate context but which presumably has some iconic aspects. The similarity between a signal and its target is thought to scaffold interpretation; all else equal, iconic signals are easier to interpret and learn than others. This of course depends on the similarities being salient and suggestive. The wiring diagrams for electric appliances are icons; there are objective similarities between diagram and object. But these are so abstract that their interpretation requires a rich cultural scaffolding. Symbols, like icons, contrast with indexes in not having their targets restricted to the here and now. Unlike icons, symbols bear no relevant resemblance to their targets; their interpretation is not scaffolded by similarity. Hence their interpretation (on the standard way of thinking) imposes the greatest cognitive demands.

We think there is something to this picture. Indeed, we think that much of the initial expansion of communication in the hominin lineage was an expansion in the use of gesture and mime, and we think this in part because gesture and mime have iconic elements (see section 2.3). Even so, we are skeptical about the Peircean framework; we propose to use an alternative sender-receiver framework to structure our account of the initial expansion of hominin communication. In this section we explain our reservations about the index-icon-symbol model, and in the next develop our alternative, while noting some recent critiques of the sender-receiver framework as a model of language evolution.

We have three main reservations about the Peircean model. First, indexes are less obviously learnable by association than a first pass through the examples suggests; and for the same reason, association is less central to their significance. For indexes typically have both an indicative and an imperative message. The vervet leopard call both notes a leopard in the vicinity and directs a specific course of action. Likewise, the mating display of the lyrebird is a claim to be a high-quality male, and invites mating. But while most leopard calls are followed (we would guess) by dashes high into the nearest tree, invitations to mate are routinely rebuffed. There is no close correlation between mating displays and subsequent mating. While presumably there is a close correlation between mating displays and readiness to mate, readiness to mate is not a salient and observable feature of the environment. The meaning of the display is not readily learned by noting its regular co-presence with a ready-to-mate male. The most obvious third-party interpretation of a lyrebird mating display is that it is an index of the presence of a female. There are many examples in which the biological meaning of a call is not given by the feature of the environment it most regularly occurs with. Many ground-feeding birds (pied stilts, for example) are notoriously skittish, primed to call and flee. Perhaps that is not surprising. If the cost of a false alarm is low, and the cost of missed danger is very high, there may be far more false than veridical alarm calls. Presumably, even in these very noisy cases, the signal raises the probability of the target state. Copulation after invitation is rare, but copulation without display is rarer still. But would probability-raising in a noisy environment suffice for associatively driven interpretation, for cognitively unsophisticated agents? Would a cognitively unsophisticated agent, not developmentally primed to attend to stilts, learn that a stilt's call was correlated with a modestly increased level of danger? For calls are correlated with many features of the local environment, not just one. In that sense, they are typically very ambiguous: the stilt's alarm call is correlated with (for example) godwits taking off.

The crucial point is that indexes do not typically become established in a population by receivers coming to notice a covariation between a call and a salient feature of the current environment (and of agents' responses to that feature). It is not correlation but coadaptation that connects the stilt call to potential danger, as indexical signals typically become established as the result of evolutionary or developmental coadaptation between producer and audience. So it is true that animal signals typically, perhaps always, direct the receiver's attention to some feature of the immediate environment. Thus they seem to be indexes. However, that is a consequence of the kinds of interactions animal signals mediate: courtship, warning, collective defense, collective attack. So the signals indicate danger or opportunity here-now. They do typically correlate with, associate with, features of the immediate world of the sender and receiver. But this follows from the actions they coordinate, as animals do not plan collective action in the future, rather than being an inherent constraint that restricts animal signals to indexes. Animal signals are index-like through constraints on animal joint action, not through constraints on their ability to use other kinds of signal.

Selection can find noisy correlations because it works with large population sizes; noise will not mask a correlation, and there are no constraints on memory. The paradigm of this coadaptation is ritualization. An action that was originally functional—a dog's baring its teeth preparing to attack—is recruited as a warning, a threat, as the audience treats it as evidence of danger and responds accordingly. In this case, the coadaptation was probably mediated at least partly through genetic change, but mutual adaptation can take place within a single generation. An infant chimp comes to signal the desire to be picked up by lifting its arms, recruiting as a signal what was once a functional part of actually being picked up. The infant lifts its arms to allow its mother to grip its torso without trapping those arms. Associative learning, in many cases, is surely at the heart of turning a cue[10] into a signal. To that extent, the idea that indexes rely on simple, widely distributed cognitive mechanisms is right. Indexes rest on associative learning capacities, which are indeed ancient and widely distributed. But this picture misses the critical role of producer-audience coadaptation in building signals. That role is central to the sender-receiver framework. In our view sender-receiver coadaptation is central to the emergence of protolanguages, and that is our main reason for preferring that framework.

Our second reservation is that, if an index is just any signal that has its target at the place and time of production, then this is a very heterogeneous category. It is true that all or most nonhuman signals are indexes, on this definition. We suspect most of these will have been established by some form of coadaptation mediated by evolution or associative learning. But many human signals are also indexes, and they are a very mixed bag. They include exclamations, blushes, and other more or less involuntary signals of arousal. But they also include road signs: a red light means stop, here, now. They include some gestures. Pointing is (almost always) pointing at something in the here, now. They include personal insignias of various kinds: jewelry that displays the wearer's wealth and status; uniforms that show rank and role; tattoos and other bodily alterations that display tribal identity. They include labels. A street sign that says "Westgarth St" labels the street that the sign is in as Westgarth Street. The phrase "Westgarth St" itself is not an index. It obviously can be used meaningfully far away in space and time from that street. But the street sign is an index. So while there is a core group of indexes that are bedrock examples of communication, even within this bedrock, the process of mutual adjustment is as important as the role of associative learning. Moreover, many other cases vary in their cognitive and social prerequisites. It is very unlikely that these appeared together as part of the hominin communication portfolio.

Third, interpreting an icon can be demanding. Icons map onto their target with the help of signal-target similarities. But the interpreter needs to work out what features of the signal are supposed to resemble the target (whatever that is), and what that resemblance is supposed to be. Ron points in the direction of a small dark object partially obscured in the tree, and then extends his arms straight out from his sides, drops them quickly straight down, raises them quickly again, and repeats a few times. He is miming flapping, indicating that he is pointing at a bird. There is a similarity between his actions and those of a bird flying, but an audience needs

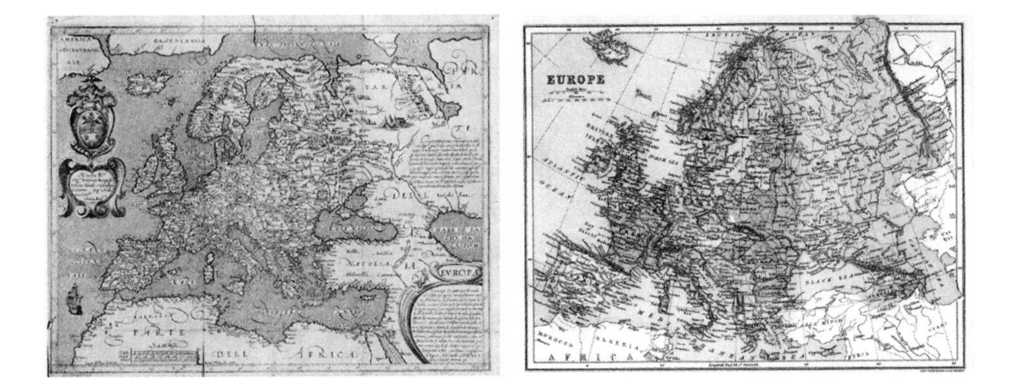

Figure 1.1
A map of Europe drawn in 1602 ("Europae" by Arnoldo di Arnoldi, Ars Electronica: CC BY-NC-ND 2.0), and a map of the same region drawn in 1923 ("Europe after WW1" by Jon Ingram, *Gresham Encyclopedia*, CC BY 2.0).

to recognize that the arm movement is a further sign, and that there is a similarity between sign and target. So long as his audience realizes that he is trying to indicate something about the environment by pointing, and his arm motions are a mime of what that something does, his communicative attempts are likely to succeed. But notice that these are rich cognitive demands (Planer and Kalkman 2019). This exemplifies on-the-fly improvisation. Other examples are more stereotyped, but they then rely on very rich shared background built by cultural learning. The iconic systems we regularly encounter are embedded in a very rich set of conventions and regularities. Maps really do have similarities with what they map, but European maps of four or five hundred years ago look very different from modern maps of the same places (and not just because the places have changed) (figure 1.1). Maps, diagrams, pictures, and other icon systems almost always have to be interpreted with the help of conventions[11] which specify what is supposed to be similar between map and target, and how it is supposed to be similar. Many Australian aboriginal cultures have strongly established traditions of storytelling scaffolded by simultaneously produced sand maps, with those maps representing the space in which the story unfolds. These maps are a mix of iconic and conventional elements (Wilkins 1997; Green 2016). Hybrid representational formats are not inventions of modernity.

In sum: the index-icon-symbol progression initially looks like a plausible picture of smooth, incremental progress in signal complexity. On close inspection, that appearance dissolves. Indexes are not a uniform category. Moreover, there looks to be a very large gap between the base case of indexes—established by mutual adjustment mediated by automatic associative learning—and even relatively simple forms of iconic mime, as in the point and flap example. Other iconic representations seem at least as demanding, relying on awareness of each other's intentions and some capacity to anticipate how others will respond to an iconic signal. Significant cultural background is often needed to interpret icons, in part because the similarities are abstract, in part because many iconic signals are hybrids, with both iconic and arbitrary elements. The placement of roads on a topographical map of the New South Wales south coast is iconic. There is a natural relation between their placement on the map and on the landscape. But the color of the roads on the map, denoting whether the road is sealed, gravel, or four-wheel-drive only, is conventional. You need to read the key to understand the significance of the colors. The icon-to-symbol transition, likewise, is not a plausible incremental step. For one thing, we have no positive characterization of a symbol; we just know that they are neither context-bound nor iconic. Symbols, too, seem to be a heterogeneous category. Vervet calls are arbitrary, and in that respect symbol-like. But they are holistic, not structured, and have fused descriptive and imperative content. Prima facie, musical notation is representational and arbitrary, and hence symbolic, but very different from words.

1.4 The Sender-Receiver Framework

So rather than develop an account of the early expansion of communication through an index-icon-symbol trichotomy, we will use the sender-receiver framework. This was originally developed by David Lewis to show how there could be conventions in communication without explicit agreement setting those conventions, so that the conventional use of signs did not presuppose the prior existence of a language (Lewis 1969). That framework was then given an evolutionary makeover by Brian Skyrms and a cluster of co-workers (Skyrms 2010). He made an even stronger point: in the right circumstances, the regular, adaptive use of signals—"signaling strategies"—can emerge and stabilize in a population without intelligent agency of any kind, let alone language. Bacteria can evolve to send and receive signals. With agents capable of learning, the simplest kind of reinforcement

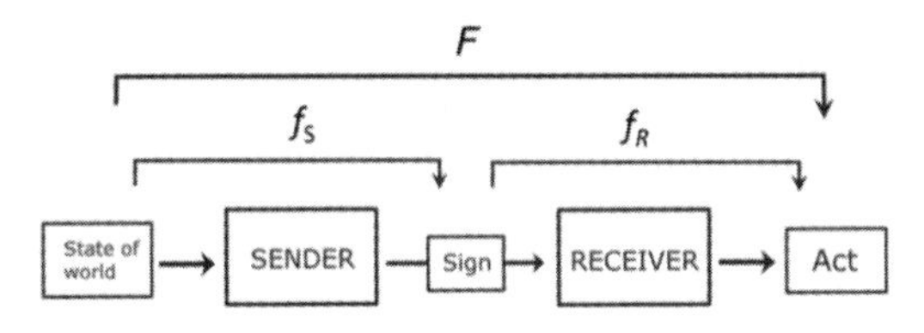

f_S: sender's rule, maps states of the world to signs.

f_R: receiver's rule, maps signs to acts.

F: the resulting mapping from states to acts.

F^*: the mapping from states to acts preferred by both agents.

Figure 1.2
The sender-receiver setup. The sender can see the state of the world but cannot act in that state except to send a signal. The receiver cannot see the state of the world but can perform one or another act depending on the signal it receives. The act that is performed in a given state yields a payoff for both sender and receiver. These payoffs shape sender and receiver strategies over time. (Figure reproduced from Godfrey-Smith 2017 with permission of author.)

learning suffices. At approximately the same time, evolutionary behavioral ecology built essentially the same framework in explaining the evolution of animal communication (for an early review, see Hauser 1996). This research tradition placed a particular emphasis on the survival of honest signaling when the interests of senders and receivers do not coincide[12] (Maynard Smith and Harper 2003; Searcy and Nowicki 2005).

Skyrms's basic model is very simple indeed. The setup consists of two asymmetrical agents, a sender and a receiver, and a world which can vary (between just two states) in ways equally relevant to the interests of both (figure 1.2). An appropriate response to that variation pays off for both. The sender can see the state of the world, cannot respond directly to it, but can act in ways visible to the receiver. The receiver can act, but can see only the sender, not the state of the world. In this ultrasimple setup, if we have a population of agents that can vary their behavior and respond appropriately to success and failure, signaling strategies will emerge and one will stabilize. The sender will produce a unique signal for each of the two world states, and the receiver will take note of that signal and produce the act that is adaptive for that state (F^* in figure 1.2).

Skyrms and his colleagues have explored the consequences of successively complicating this model by relaxing its extreme idealization: increasing the number of world states; introducing error; allowing for some forms of conflict of interest; introducing mismatches between the number of available signals and the number of relevant world states; exploring multiple sender–multiple receiver networks. The general finding is that signaling, though not always maximally informative signaling, is robust, with a strong tendency to emerge by adjustment of sender to receiver and vice versa. The same can be said of act-act variants on the above model, strategic situations where the point of signaling is to coordinate the acts of agents with one another, as in a dance, rather than to produce a collective response to a specific state of the world.[13] In contrast to the index-icon-symbol trichotomy, the simplicity and explicitness of the sender-receiver framework has lent itself to formal modeling, and that has been the crucial tool of Skyrms and the others. This further modeling work is more realistic than the elementary scenario, but even so these models simplify in ways that we will need to keep in mind when considering the emergence of protolanguage.

We think two idealizations in this modeling need particular attention. One is treating one-on-one communication as the most basic form. We think instead that the earliest forms of expanded hominin communication were established in small groups. There were always eavesdroppers and alternative sources of information. That matters, because the problem of manipulation and deception has loomed large in the evolution-of-communication literature, yet both the opportunity and the temptation to deceive are significantly suppressed in multi-sender, multi-receiver contexts. With other senders present, it is much harder to control the flow of deceptive misinformation to a target receiver. With multiple receivers, it is much harder to predict the consequences of attempts at deception, whether successful or not. The problem of explaining honest communication in a world of conflicting interest is exacerbated by thinking of private talk as the basic form of communication. A second idealization is treating payoffs one by one. Sender-receiver models are typically atomistic; signals are stabilized one by one, by the sender and receiver payoffs for that signal. The benefits that stabilize the system accumulate from each atomic interaction. But once even a modest protolanguage has evolved, this representation is not realistic. Many interactions, perhaps most, do not issue in any identifiable action. Moreover the basic atom of anything like language is a many-way conversational interaction, often quite extended, perhaps with varying participants, not a unidirectional flow of information or instruction from speaker to audience.

The shift from payoffs as captured in sender-receiver models to payoffs dependent on possessing a more general communicative capacity probably happened quite early in the expansion of hominin communication. On average, communication had to pay its way. But we suspect that at some fairly early stage in this process, what paid was a more general capacity to signal and interpret, not specific signal-response pairs. While the modeling results in the sender-receiver framework are very suggestive, the simplifications make it difficult to apply the models to real cases. Nevertheless, the broader message of these models is that communication is, first and foremost, a dynamic process of sender-receiver coadaptation. That point is not compromised by the models' idealizations, for it is not an inherently formal idea.[14]

Before closing this section, we should mention one other framework. Thom Scott-Phillips suggests that our favored sender-receiver framework is a good general model of animal communication but not of most human communication, including language. Building on the work of Dan Sperber and others, he distinguishes between codes and ostensive-inferential communicative systems (Scott-Phillips 2015). All or most animal communication is code-based. Codes have a menu of signals with fixed meanings. The menu is often small: a small set of calls and responses. Sometimes they are limited but continuous, as when the amplitude of the call signals the urgency of a threat. Occasionally they are structured, as with the honeybee system. In principle, a code could be infinite; it might feature some form of recursion or iteration. But all codes feature a fixed stock of basic signals. Sender-receiver models are natural ways of representing these codes. Humans certainly use codes. At one stage, chess was played by correspondence, and there were fully explicit codes for representing successive moves. Likewise, bidding conventions in bridge are codes. But these are restricted, special-purpose systems. Often human communication depends on an interpreter recognizing that a sender is attempting to communicate. On the basis of that recognition, together with common ground and general contextual information, the audience works out what the speaker intends to communicate, usually so swiftly that the inference is both rapid and tacit. Kim catches a friend's eye at a dull party, and thus swiftly but unobtrusively sends the message that it is time to go. This is as true of language as it is of other forms of communication, as we noted in section 1.1 in discussing pragmatics and the distinction between what was literally said and what was conveyed. Scott-Phillips suggests very plausibly that ostensive-inferential communication is unbounded. There is no fixed stock of basic signals; in the right context anything can be recruited as a signal. Moreover, the messages we send through language outrun their literal meaning. Much more controversially, he argues that there can be no incremental transition between codes and ostensive-inferential communication. Codes, according to him, depend on the cognitive machinery of association; ostensive-inferential communication depends on rich theory-of-mind capacities.

There are important insights in this renewed emphasis on the pragmatics of language and the duality of anticipation and interpretation. But we have two important reservations about this framework. In chapter 4 we will take issue with the sharp distinction between codes and ostensive-inferential communication, showing that systems more flexible than codes but less flexible than our contemporary capacities are available to agents with the capacity to imitate; with some, though limited, theory-of-mind capacities; and with some capacities for conditional reasoning. Second, we have a more general methodological reservation about this program and a number of others in the evolution-of-language industry (for example, those relying on the notion of joint or shared intentionality).

Scott-Phillips and his predecessors frame their account of ostensive-inferential communication using the categories of folk psychology: of beliefs and intentions. More specifically, on their view, when humans communicate, they are supposedly doing so through having "higher-order" intentional states: speakers intend hearers to form beliefs about the speaker's own intentions and beliefs. In the terminology that has been co-opted for this line of thought, a first-order intentional system is an agent capable of having beliefs (and desires); for example, the belief that a tiger is over the next hill. But such an agent does not have beliefs about beliefs. Second-order intentional systems have beliefs and intentions about the beliefs and intentions of other agents (and about their own). Third-order intentional systems have beliefs, intentions, and the like about beliefs that are themselves beliefs or intentions about beliefs, intentions, and the like. Ron can hope$_3$ that Kim believes$_2$ that Ron intends$_1$ to finish this chapter tonight. According to this framework, ostensive-inferential communication is possible only between agents who are higher-order intentional systems. Moreover, we supposedly are very-high-order intentional systems because we can reasonably reliably follow narratives in which Dave imagines that Amanda believes that Barry intended her to think he did not want to come to the party. Along similar lines, Clive Gamble, Robin Dunbar, and John Gowlett attempt to pair relative neocortical volume in living and extinct hominins with their order of intentionality (Gamble, Gowlett, and Dunbar 2014).

This way of framing the issues creates two problems. The first is that cognitive psychology seems to have all but dropped out of the picture (an analogous point is made by Celia Heyes about theories of cultural evolution; Heyes 2018). For what is the relation between an agent's intentional states—her beliefs and preferences—and the cognitive mechanisms that causally explain her actions? There is a line of thought, most famously defended by Jerry Fodor, that folk psychology is a *form* of cognitive psychology. When we correctly attribute beliefs and preferences to an agent in explaining, say, why she left the pub, we identify representational states of that agent's central processor that interact in ways that drive the motor commands that took her out of the pub (Fodor 1975; Fodor 1983). But this is a very controversial view of the relation between folk psychological states and cognitive mechanisms. For example, on Dan Dennett's view, the relationship between folk psychological interpretation and cognitive mechanism is very indirect. As he sees it, a folk psychological interpretation captures reasonably stable patterns in another agent's actions. We use those patterns to predict, coordinate, and intervene, usually with a good deal of success, even though the patterns are noisy, and even though we typically know little of the cognitive mechanisms that explain these patterns, their stability, and the exceptions to them (Dennett 1975; Dennett 1991). If any view like Dennett's is right, cognitive psychology has been sidelined by framing the evolution of language through the lens of folk psychology. To the extent that folk psychology is a gadget that humans have collectively assembled to interpret one another, it seems unlikely that our own folk psychology is the best framework for describing the mechanisms of minds significantly different from those of contemporary humans. Yet the minds of great apes and earlier hominins almost certainly were significantly different from our own. This leads to our next point.

Second, and perhaps more importantly still, this framework gives us little handle on intermediate cognitive engines: cognitive systems that are not just bundles of reflexes and associations, but which lack the sophistication of full intentional psychology. In discussing great ape communication (section 2.1), we will see examples like this: behaviors that seem too nuanced and sophisticated to be treated as mere reflexes or conditioned responses, but too limited and inflexible to be fully voluntary. Moreover, while the framework does make distinctions within the class of intentional systems, we wonder whether these are the best ones to draw. Are orders of intentionality genuine architectural differences between cognitive engines, or are they (as introspection might suggest) strongly influenced by the concentration, context, and attentional focus the agent brings to the challenge? There are also differences between intentional systems that are orthogonal to orders of intentionality. We can indeed intend to go to the pub, believe we are going to the pub, and prefer to go to the pub. But we can also imagine or pretend we are going to the pub. These capacities may well be important to the narrative uses of language, and may be more of an architectural difference than higher orders of intentionality. These considerations make us suspect that it is unwise to frame the evolution of communication through the lens of folk psychology and orders of intentionality. After all, folk psychology is a set of conceptual tools that has emerged in and for our social lives. It allows us, with imperfect but still remarkable reliability, to interpret each other's actions and to anticipate and influence future actions. It was not designed to specify the cognitive architectures of great apes or other hominins.

1.5 Empirical Constraints and Language Evolution

Famously, explaining the emergence of language is made difficult by evidential considerations. As those working on the evolution of language endlessly grumble, language does not fossilize. It is true that in favorable circumstances, there is fossil evidence of voluntary control of complex vocalization, but we have already noted that vocal control is neither necessary nor sufficient for language. There have also been some recent suggestions of genetic fossils that give us a direct historical signal about the antiquity of language (Dediu and Levinson 2013; Levinson and Dediu 2018). For there is now direct evidence about the genome of two extinct human species: Neanderthals and Denisovans. These genomes can be compared to each other and to ours, enabling us to make reasonable estimates of the genome of *Homo heidelbergensis*, the presumptive common ancestor of *sapiens*, Neanderthals, and Denisovans. To the extent that we can identify genes especially relevant to our possession of language, we can estimate the presence of these genes in our sibling species, and in our common ancestor. Dan Dediu and Stephen Levinson (2013) have recently argued on these grounds that AMHs, Neanderthals, and Denisovans all had language by inheritance from our common ancestor. While this line of thought is innovative, these genetic fossils at best show that these other human species had the genetic resources needed for language. That is by no means the same thing as having language. Our deep-time ancestors very likely had the genetic resources needed for formal quantitative reasoning, but without the cultural invention of numerals and a number line, those resources could not be exploited. The same may be true of language. The central role of cultural learning in the construction and transmission of language (qua social phenomenon) is enough to show that the use of language depends on cultural scaffolds, not just appropriate genetic potential. In any case, this line of evidence cannot tell us anything about the incremental sequence of language evolution. Instead, our argument will mostly rely on archaeological phenomena and their implications, though in chapter 5 we will make some appeal to neuroscience. (We discuss genetics some more in chapter 6, though genetic evidence functions there more as a consistency check on our ideas, as opposed to positive evidence in support of those ideas.)

Our strategy is to nest an incremental view of the emergence of language within an equally incremental view of the evolution of human social life: in particular of the extent and nature of cooperation, and the extent and nature of social learning. This more general framework is developed in Sterelny (2012a); Sterelny (2021). Communication (typically) is itself a form of cooperation, but it is also an essential adjunct for the coordination and negotiation needed for other forms of cooperation to be stable and productive. For example, some forms of cooperation are stable only if reputation (knowledge of the past social actions of others) is tracked reliably and is part of common knowledge. So evidence of the presence of such forms of cooperation is evidence of quite sophisticated communication. Hence for us, the forms of cooperation carry critical information about the nature and extent of communication. The nature of cooperation changed over time, for while the evolution of human social life was characterized by the expanding role of cooperation, that expansion was not seamless. In our view, there were two cooperation transitions in ancient hominin life. One took our lineage from great ape levels of cooperation to coordinating forager bands specializing in team-based hunting of large game (Bunn 2007; Bunn and Pickering 2010; Pickering 2013; Bunn and Gurtov 2014; Sterelny 2014). In the second, the life of mutualist foragers transitioned to reciprocation-based, small-party foraging around 100k years ago (Sterelny 2014; Sterelny 2021). These two forms of economic organization imposed quite different demands on communication. In particular, we shall argue that this second economic and social transformation had profound consequences for the communicative needs of human foragers, pushing our late Pleistocene ancestors to, or close to, something we could call language.

In summary, then, humans evolved as social, increasingly cooperative, technique-and-technology-enhanced foragers over several million years. Changes in their technical capacities, and in the targets and techniques of their foraging, left material signatures. To some reasonable extent, we can track where hominins lived, the resources they exploited, and the technologies they used in exploiting those resources. Those aspects of hominin lives left traces. In turn, technique, technology, and changes in technique and technology interacted with the economic and social organization of forager lifeways, and hence carry information about those lifeways. Technique and technology, and changes in technique and technology, also carry crucial information about the cognitive capacities of those ancient hominins. Material technology can signal the possession of cognitive capacities that make new forms of communication available.

There is nothing original in the claim that proposals about the timing and shape of the evolution of language should be tested against the material record of hominin evolution. However, most attempts to tie the evolution of language to the paleoanthropological record have looked for a specific signal of the arrival of language: *a language signature*. For example, the regular use of material symbols is often seen as the signature of language (see, for instance, Tattersall 2016); so, too, long-distance trade networks (Marwick 2003), or systematic teaching (Laland 2017). We are not in the hunt for a specific signature of language, or even of the various versions of protolanguage. Rather, we aim to integrate information about

(i) different foraging economies and the communication and coordination demands those economies impose; and
(ii) the cognitive capacities implied by the manufacture, use, and social transmission of different technological suites.

In the rest of the book, a particular pattern of inference will become familiar. We will try to show, for instance, that those who made and used

Acheulian stone tools must have had, given the task demands that confronted them, a cluster of cognitive abilities that made certain aspects of language-like communication available to them: say, the ability to coin new signs with reasonable success. Given those hominins' cognitive capacities as shown by their technology, these communicative capacities were within reach. Competences that help make language possible are manifest in other aspects of hominin lifeways. We then argue, in parallel, that their social environment would have selected for those communicative capacities. In short, these capacities would have been both useful and potentially available. Our standard pattern of inference is to use the material record to identify the availability of cognitive and social resources for particular communicative capacities, and to use that same record to identify communicative needs that select for those capacities. If capacities were both available and useful, they probably had them.

In turn, this commits us to the view that adaptive phenotypic plasticity has played a central role in the evolution of hominin social life in general, and of hominin communication in particular. In her magnum opus, Mary Jane West-Eberhard describes this as the view that genes are the followers, not the leaders, in evolution (West-Eberhard 2003). A behavioral innovation is successful, and its success stabilizes it through various forms of learning. Once it stabilizes, if it is important enough, and persistent enough, selection will favor genetic variants that acquire the behavior more rapidly or reliably. Likewise it will favor variants that elaborate and/or find new uses for the behavior. Genetic accommodation, as West-Eberhard calls it, is by no means inevitable. But in our view, hominin evolution was often shaped by feedback between individual and cultural learning on one side and genetic change on the other. Individual and cultural learning triggered change in the social environment, and those changes then selected for genetic change.

So one continuing theme in our discussion is this argument pattern linking ecology, technology, and communication. A second continuing theme concerns changes in social learning. For the most rudimentary protolanguages and their less rudimentary successors are transmitted from one generation to the next by social learning, and the richer protolanguages are difficult social learning targets (see section 2.2). So we will need to identify the archaeological and paleoanthropological signatures of extensive and accurate social learning, the kind needed for language, or for an expressively rich protolanguage. Without reliable and accurate social learning, innovations are likely to disappear with the innovator without leaving a trace. So if there are increasing archaeological traces of innovation, that is evidence of enhanced social learning. We will suggest that this signature is seen quite late in hominin evolution. Between 800 kya and 500 kya, there is some, perhaps rather unstable evidence of innovation and hence of improving social learning. Over that period of the mid-Pleistocene, a cluster of very important technical innovations appears in the record, though not with great stability and persistence. These include the domestication of fire (Alperson-Afil, Richter, et al. 2007; Wrangham 2017), and an up-scaled version of Acheulian technologies at about 800 kya. At around 500 kya, we find much more elaborate forms of prepared-core techniques, giving artisans much more control over the form of the flakes struck from the core, and the first unambiguous signs of composite tools.[15] But there was a marked and further increase in the rate of innovation perhaps somewhere between 200 kya and 100 kya, with increasing acceleration over the last 100k years, though still with a good deal of instability (Sterelny 2011; Sterelny 2012a). The overall pattern seems to have been one of very slow change from the first stone tools at 3.3 mya, then a gradual but by no means smooth increase in the rate of innovation and diversification from about 800 kya. Much more happened in the final hundred thousand years of the Pleistocene than in the preceding 3.2 million years of stonecraft. This narrative is probably oversimplified through its excessive focus on stone technology. The spread of the erectines though much of the Old World, and the harvest of large game, suggest the social transmission of a good deal of information about the local environment and its natural history. Even so, there is a marked though patchy and unstable acceleration of change in the last 200k years. We (and many others) see this as a signature of increasingly reliable cultural learning (Henrich 2004; Powell, Shennan, et al. 2009). We return to this in chapter 7. The details are complex, messy, and controversial, but two big-picture points are clear. First, a rich protolanguage is stable only if most of a community uses it, and such a protolanguage can be acquired only by agents who are good at learning from others. Second, there is no archaeological signature of reliable, accurate, large-volume social learning before (say) 800 kya.

We will finish this chapter with a more general moral. The evidential streams from the historical sciences enable us to form, in an admittedly fragmentary and fallible way, pictures of the differing social worlds of long-vanished hominins, and of the ways those worlds required communicative capacities ancestral to language. We must be cautious here: it is alarmingly easy to imagine we know more about the lives of these vanished hominins than we really do. Even so, while we rarely have direct evidence about ancient communication systems, we regularly have some information about ancient groups' more general capacities to accumulate and transmit information. We can use evidence of how these groups lived, where they lived (and hence their environmental tolerances), what they made, and what they ate to probe the communicative and social learning prerequisites of their lifeways. This is the inferential engine of the rest of the book.

In the next chapter, we will identify the launching pad from which hominin communication expanded, combining comparative data about great ape capacities with archaeological data about hominins on the verge of adopting lifeways dependent on stone technologies and cooperative scavenging and hunting, as well as giving our account of the capacities needed for full language, thus bookending the launch and the terminus of a lineage explanation.

2 Toward Protolanguage

2.1 The Great Ape Baseline

In the last chapter, we explained and defended the framework assumptions that structure our view of the evolution of language. In this chapter, we begin to build that view. In building a lineage explanation of the evolution of language, a crucial first step is establishing the baseline. What were the communicative and cognitive competences of early hominins? What were the capacities that formed the foundation for expanded communication, and ultimately for language? There is no perfect way of answering that question, but the best option is to look at living great apes. Hominins diverged from the chimp lineage perhaps 7 mya, and the combined hominin/*Pan* lineage diverged from the gorillas and the orangutans earlier still. These lineages did not freeze at the point of divergence. If our split from the chimps was 7 million years ago, we have been separated from their lineage by 14 million years of diversifying evolutionary history. That said, the communicative abilities of the living great apes, and the cognitive skills that support those abilities, appear broadly similar. Indeed, according to Byrne, Cartmill, et al. (2017), there is considerable overlap in the gestural repertoires of the great apes. In particular, this team has found that both chimps and bonobos have similar maps from gestural forms to messages (typically, requests). These facts have led Byrne et al. to hypothesize that great apes more generally might employ a similar set of such mappings. However, even if their specific repertoires turn out to vary quite significantly, as they probably do, their general capacities seem quite similar, and it is these general capacities we suspect were shared by early hominins. For it seems safe to assume that the great apes have not sharply diverged from each other in this regard, and that in turn suggests that the hominins have diverged from a shared great ape inheritance. What was that inheritance like?

Let's begin with vocal communication. Until recently, the collective wisdom about great ape vocal communication ran as follows (see, e.g., Tomasello 2008). The great ape vocal repertoire—of chimps, say—is small. There is little sign of local variation, and very little sign of their vocal repertoire being open to modification by learning. Chimps (and other great apes) are unable to add new calls, and are unable to adapt existing calls to new functions. Calling is probably involuntary, not under top-down control. The calls of chimps and other great apes are largely or entirely a consequence of various forms of arousal: fear, sexual excitement, social anxiety, greed, anger. A few human vocalizations are like that: exclamations of surprise, fear, pain, pleasure; perhaps laughter. What for humans is a tiny fraction of their vocal repertoire is the whole thing for great apes.

In short: great ape vocal communication was seen as limited, inflexible, and driven by ancient, widely shared neural mechanisms. But great apes communicate by gesture, not just voice, and to many theorists their gestural communication has seemed strikingly different. The number of great ape gestures (including those used by chimps) is very large. Moreover, the menu of the gestures available to an individual ape at a time is shaped by experience.

Gesture is voluntary or intentional. Apes gesture in ways that are contextually appropriate, and change their communicative strategies adaptively in response to their audiences' responses. Partially as a consequence, a significant group within the evolution of language research community have argued that within the hominin lineage, gesture led the expansion of communication beyond great ape baselines (Arbib, Liebal, and Pika 2008; Tomasello 2008; Corballis 2009). We also belong to this group. The perceived contrast between rigid, reflex-like great ape vocalization and flexible, intelligent great ape gesturing has been one factor driving this view. Nonetheless, recent work makes it very likely that the contrast between gestures and calls has been overdrawn. In what follows we will discuss this revised view of great ape (and primate) vocalization. We still think that the evidence shows a very significant difference between the two communicative modalities, and so we still take gesture to be central to the early expansion of hominin communication, though very likely hominin communication was always multimodal, with vocalization also changing as hominin social life became more complex and more cooperative.

In a series of papers, Catherine Crockford, Klaus Zuberbühler, and their co-workers have argued that chimps show considerable flexibility in their warning calls; indeed, they think there is evidence for voluntary, intentional control of these calls (Crockford, Wittig, et al. 2012; Schel, Townsend, et al. 2013; Crockford, Wittig, et al. 2017). The more modest claim of situational flexibility is well-supported; the claim for intentional, voluntary control is intrinsically more difficult to defend. Their experimental subjects are wild but habituated chimps. The researchers' technique is to anticipate the trail along which a foraging chimp will move, and to place near that trail, not too obviously, a model of a venomous snake (in most of their experiments). Chimps utter a range of warning calls, and the aim of the experiments is to understand the effects on the sender of variations in the ecological and social context. Do chimps vocalize only if there is an audience? Is the vocalization affected by the character of the audience; that is, do they warn more, or more urgently, if the party following is composed of kith and kin? If calling is driven by simple arousal, we would expect calling to diminish or stop as the chimp moved away (though snakes are really dangerous only when unseen). If, instead, calling is focused on the receivers, we would expect a different stopping rule. The chimp will stop calling when the receivers move into safety, or show that they have already seen the snake (for example, by beginning to alarm-call themselves).

As always, the data are noisy, but in general they support the view that calling is contextually sensitive rather than simple and reflex-like (for a review and diagnosis of the noisiness of the data, see Crockford 2019). The presence and character of an audience usually matter (we will discuss an exception shortly). The stopping rule is receiver-focused, and not just because the caller stops in response to the audience's situation. For the caller also actively monitors the situation of the audience, and alternates gaze between the audience and the model snake. Indeed, in one experiment, in a significant fraction of the cases calling was supplemented by the discoverer of the snake placing himself (or herself) close enough to the snake to ease the audience's task of finding the snake (for the snakes are well camouflaged and can be hard to see even if you know they are somewhere around). As the authors recognize, experiments of this kind cannot decisively rule out the possibility that calling is a response to arousal. Perhaps the presence of kin and allies increases the chimp's anxiety, causing her to call more in such contexts. But the more nuanced the response to circumstances, and more contextual variables to which calling is sensitive, the less plausible such deflationary explanations become.

What of the view that calling is intentional, voluntary? Zuberbühler and his colleagues point out that there is a fairly standard way of operationalizing this notion (though some of the most telling criteria are intrinsically restricted to visual signals). Given this operationalization, they argue that at least some chimp warning calls satisfy several of the usual criteria for intentional action: actions driven by the beliefs and preferences of the vocalizing agent (Schel, Townsend, et al. 2013). The criteria are:

Social use	The signal is produced only when there is an audience.
Social use	The signal is sensitive to the character of the audience.
Social use	The signal is sensitive to the response of the audience.
Sensitivity to attention	Visual signals are chosen only if the sender is in the receiver's visual field.
Manipulating attention	Before a visual signal is produced, the sender engages in attention-getting behavior.
Audience checking and gaze alternation	The sender checks the audience and visually orients toward them; when the signal is about some third entity, the sender alternates gaze between receiver and referent.
Persistence/ elaboration	The sender persists in signaling (perhaps using different signals) until the communicative goal is met.

Chimps do not produce distinct calls for distinct threats, but they do produce three alarms: a soft hoot, an alert hoot, and a wah bark. The researchers' crucial claims are that (i) alerts and barks are made only in the presence of an audience; (ii) their mode is sensitive to the character of the audience; and (iii) the stopping rule is other-directed, as discussed above. Calls fade as the presumptive audience is out of danger, and there is evidence of gaze alternation between snake and audience, and of checking where the audience is. One might see this as evidence that alerts and barks are deliberate, intentional acts, and that chimps are at least first-order intentional systems. We see this experiment in a different way: it shows how difficult it is to show intentional control and perhaps how unclear that notion is. For one thing, some aspects of the sensitivity of calls to social context are puzzling. For while it makes sense to care more about kin and allies, and to call more when they are the receivers, the chimp that first notices the snake also calls more when followed by a dominant chimp, and they certainly have no reason to protect a dominant. On the other hand, the proximity of a dominant might increase arousal. Soft hoots are even more of a puzzle. They show a strange mix; there is no evidence of their social use on any of those criteria. Soft hoots are sometimes produced without an audience, and independently of the kind of audience (if there is one) and their response. Yet when there was an audience and a chimp hooted, there was evidence of gaze alternation and audience checking. What should we say about such mixed cases? This is just one aspect of one experiment, but we think it shows how difficult it is to shoehorn nonhuman agents into the distinctions between zero-order, first-order, and second-order intentional systems. That said, the experiments certainly suggest considerable flexibility and context sensitivity in the production of some calls.

We have so far focused, with Crockford and Zuberbühler, on the production of calls, and the upshot is that the control of such calls is more contextually sensitive than had previously been thought. In a series of papers, Dorothy Cheney and Richard Seyfarth focus on *reception*, and argue that primate calling is an unexpectedly, cryptically rich communication system (Seyfarth and Cheney 2010; Seyfarth and Cheney 2014; Seyfarth and Cheney 2018a; Seyfarth and Cheney 2018b). They draw mostly from their work on baboons, but it is clear that they intend their thesis to apply widely across primate vocal communication. They distinguish between the production, use, and interpretation of calls. They agree that primate call production is highly constrained: primates choose from a very small menu of signal types, and that menu is difficult or impossible to alter. In broad agreement with Crockford and Zuberbühler, they argue that primates have some control over when to vocalize and the manner of their vocalization: they do choose whether and how to call. But their central claim is that primates understand calls through their quite rich and conceptualized understanding of their social world.[1] Thus baboons interpret signals in the light of information about the signaler, of their past histories of interaction, and of information about the social and physical context. For example, baboons interpret both threat and reconciliation grunts in terms of their interactions not just with the senders, but also with the kin of the senders. If you have been a target of aggression from a higher-ranked individual, a threat grunt from a member of the same matriline is (rightly) treated as more dangerous. They also show these interpretive capacities as bystanders: for example by heightened alertness when they hear dominance and submission grunts that are inconsistent with the background social hierarchy. That is especially true when that surprising interaction is between matrilines rather than within a matriline, as that would be information about a major change in their social world.[2] In short baboons, and by inference many primates, have very considerable social intelligence, and they use this intelligence in their interpretation of incoming signals.

Cheney and Seyfarth leverage these claims about social intelligence and its use to make a very striking claim about the character of baboon vocal communication. On their view the system is cryptically rich. It is discrete, combinatorial, rule-governed, and open-ended. It is discrete and combinatorial because the receiver can extract from a call (i) the caller's identity; (ii) relative rank and kinship affiliation; (iii) the character of a social interaction: threat, submission, reconciliation; and (iv) often recent interaction history, as with reconciliation. In a group of 80 or so, that means a very large number of distinct messages can be sent and recognized. Discrete, additive chunks of information are built into these apparently simple signals. It is rule-governed because there is a clear mapping between call and call function. It is open-ended because the baboons in the group learn to recognize the calls of those being born into, or migrating into, a group.

There is a rat to be smelt here. On this reasoning, 'I' would have many more meanings in a large group than a small one. But 'I' in fact has only one meaning, roughly 'the speaker of this utterance'. Likewise, when a baby cries, its parents usually recognize that the baby is theirs. But the cry is not a signal of specific identity but of unhappiness. Cheney and Seyfarth seem to be going wrong in two ways here. First, they are failing to distinguish between signal and cue. Not everything about a vocal performance is a vocal signal. Sterelny's speech carries information about his age, sex, and health, but almost certainly as a side effect of genuine adaptations for its signaling function.[3] Likewise, the specific identity conveyed by a threat grunt is a cue rather than a signal. The same is true of information about kinship and rank, even if that information is carried by the sender's vocalizing. Moreover, Cheney and Seyfarth have exaggerated the amount of information a typical signal carries. It is true that when an eavesdropper hears A threatening B, he/she knows the identity and relative rank of the two individuals. But that is usually part of the bystander's background knowledge. Rank is learned from the signal only when that signal indicates a dominance reversal. Even then, while it does indeed carry information about that reversal to third parties, it is not adapted to do so. These are cues to the eavesdroppers. We are then skeptical of the idea that these primate call systems are cryptically rich, despite their small and fixed menu of signals. Instead, we read Cheney and Seyfarth's analysis in another way. The expansion of the signal repertoire was facilitated by standing interpretive capacities; the latter were not a bottleneck. Once the signal production repertoire began to expand, significant interpretive capacities were already in place in the primate lineage.[4] For primates are adapted to use cues, not just signals, to harvest information about their social environment.

Summarizing these considerations, we still think that expansion began in the gestural mode. We accept that call production is more flexible than had previously been supposed. But no new evidence challenges the idea that call repertoires are small, sharply fixed, and very resistant to modification through experience, and we infer that the same was true of the last common hominin/*Pan* ancestor. This is not true of the gestural repertoire, and that makes it a much more readily modifiable resource once there is selection for expanded communication. We sketch that idea in section 2.3. Our next step, though, is to identify the core cognitive and social capacities on which current languages depend. We specify those capacities below. We think that all or most of them existed in some form in early hominins, indeed in the last common ancestor of the hominin and *Pan* lineages. If that is right, a lineage explanation of the evolution of language identifies a set of initially fairly independent cognitive capacities whose integration, coevolution, and enrichment explain the emergence of language. This claim is methodologically critical. For as we saw in section 1.5, our view of how to build an empirically constrained theory of the emergence of language depends on the view that capacities that make enhanced communication possible also support other behaviors, including those that leave more direct archaeological signatures. So the first two sections of this chapter partially specify the two endpoints of the lineage from the earliest hominins to living humans. We now move to the second endpoint.

2.2 The Language Mosaic

In this section we sketch the main cognitive mechanisms on which language relies. Our main claim is that all or most were present in rudimentary form in early hominins whose communicative capacities were roughly similar to those of living great apes.

Short-term memory, online-processing capacities, and control. Of course, primates have good short-term memories and good online processing capacities (see for example the baboon experiments described in Claidiere, Smith, et al. 2014, where baboons learn to recognize, remember, and exploit quite complex and arbitrary patterns). But language as we know it imposes extreme demands on these cognitive skills, especially given the fact that we are often engaged in nonlinguistic activities in parallel with conversation, sometimes of a demanding kind. For this reason, evolving language required an impressive expansion of control. When language is a tool for coordination in the here and now, agents have to divide attention between the demands of communication itself and the demands of the situation in play. Even when conversation is not a tool for managing coordination, we often engage in activities in parallel while talking, activities which require some attention and monitoring. But expanded control is also required because language requires the production of precise, elaborate, and error-intolerant sequences. Small differences in sound or shape can map onto big differences in upshot.

But even setting aside these issues of control and the division of cognitive work, the short-term memory demands of language are imposing. Items in the speech stream have to be recognized and retained in memory long enough for phrases, constituents composed from lexical items, to be identified, and for their relations to one another to be tracked. Think how difficult word identification is, not just because the speech stream is in fact continuous, but also because homonyms, especially those involving names, are so very common. It is not enough to identify the sequence P-E-T-E-R as a separate word and name: the listener needs to recognize which Peter is being named. Moreover, recovering who did what to whom, when, and in what circumstances often involves stitching together as a single constituent elements separated in the speech stream. Of course it is typically in a speaker's interest to be understood, and in any case listening is a fallible skill. Misfires do happen. Even so, considering the difficulty of the task and the time pressures involved, conversation exhibits impressive short-term memory capacities and even more impressive processing speed.

Semantic memory. Contemporary language also depends on good semantic memory. Languages vary in the size of their vocabularies, perhaps by an order of magnitude (Henrich 2016). World languages have huge vocabularies, though speakers only control a subset of the community's lexical resources. But even so, speakers typically use thousands of words. Moreover, these numbers exclude proper names, adding a further load, exacerbated, as noted above, by their ambiguity. The extent of this burden is somewhat controversial, for there is much dispute about what a speaker knows in virtue of having a word in their individual lexicon. In some versions of generative grammar, information in the lexicon tagged to particular lexical items does a lot of the work of preventing overgeneration.[5] Likewise, on some views of semantics, understanding a word involves knowing something like a definition (Jackson 2010). But on any view, having the standing capacity to use any of thousands of words, usually effortlessly, is a far from trivial feat of memory. For words are not just recognized. They are recalled, and often in the absence of their referent.

Moreover, semantic memory is a potential selective driver of the elaboration of communication. The more information (or misinformation) an agent remembers, the more they have to contribute to the informational commons, on the assumption that there is an informational gradient across community members. If the community always moved together through their environment, and slept close to one another, as some baboon groups do, there might not be much of an information gradient. The same may well be true of gorilla family parties. Mostly, what one agent knows, the others will also know. But the two chimp species have a fission/fusion social organization. After resting and sleeping in close proximity, chimps disperse into their territory in small groups, congregating again at night. Over much of their evolutionary history, hominins probably had a somewhat similar social organization, and this does generate useful informational differences between community members. Difference is important, but so too is overlap. As Michael Tomasello in particular has emphasized, many conversations are intelligible only because of joint attention—all parties are aware of a common scene, and know that about one another—and common knowledge (Tomasello 2008; Tomasello 2014). Communication thus depends on a balance between unique and shared information, and we shall suggest that this balance shifted over time in ways that mattered to the emergence of language. Indeed, the effects of diminishing common ground on linguistic communication have been observed in real time at schools for the deaf. In one such school, with the arrival of growing numbers of children, and from farther away, common ground was strained, resulting in greater lexical and structural richness on the part of the children's sign language (see, e.g., Meir, Sandler, et al. 2010). We revisit this issue in chapter 7, where we explore the consequences of more spatially and temporally dispersed forms of foraging, Clive Gamble's "release from proximity" (Gamble 1998). We expect that there was positive feedback between (proto)language and semantic memory: an expansion of semantic memory selected for expanded protolanguage, and protolanguage, as it became more elaborate, selected for richer and more efficiently organized memory, including remembering what you have been told and from whom, not just remembering what you have personally experienced.

Mental models. Actions typically succeed or fail depending on their causal consequences, so agents, in considering their alternatives, need to be able to estimate those consequences in advance. Mental models make this possible. For models do not merely register the objects and properties present in a situation (real or imagined). They also represent the causal dependencies among them. This supports our abilities to reason predictively, retrodictively, and counterfactually (Christensen and Michael 2015). This is obviously important in physical activities that are not stereotyped, like making a handaxe ("shall I strike the core *here* or *there* now?"). But it is equally important in the social world, including coining new signals ("if I do *this*, will he do *that*?"). As a consequence, we think an expandable protolanguage demanded enhanced causal reasoning. To expand their protolanguage, our ancestors needed an enhanced ability to reason about the likely effects of their own novel signals. Once a new communicative device (a word, a way of organizing information) was coined and caught on, the cognitive demands of learning and using that device were significantly reduced. A child might simply observe that when others perform a certain kind of gesture—say, pounding one's fist into one's palm—they tend to receive nuts from their neighbor. And so the child adds this action to his own behavioral repertoire. That is much simpler than inventing this signal. Except perhaps for a few lucky accidents that caught on, invention would have involved prediction and inference on the part of communicators, though no doubt with many unsuccessful predictions and failed signals. (More on this in section 4.2.2.) Of course, once language is up and running, this flexibility is scaffolded by language itself: "Let's call her 'Midnight'," and Midnight is so-named. In the earlier versions of protolanguage, before it acquired any metalinguistic apparatus, adding new items cannot have been so easy, and must have mostly depended on causal hunches coming off.

Theory of mind. We noted in section 1.1 the important distinction between the regular, conventional meaning of a message and what the speaker intends, often successfully, to convey by that message. Dan Sperber, Michael Tomasello, and most recently and explicitly Thom Scott-Phillips have taken this distinction to suggest a fundamental and discontinuous distinction between two kinds of communication systems: code-based systems, exemplified by the vervets' three distinct warning calls for leopard, eagle, and snake, and an "ostensive-inferential" system that we use (see especially Scott-Phillips 2015). Code-based systems constrain the sender to pick from a fixed menu of signals (recall), a set repertoire of possible signals. The audience's sole job is to recognize the signal, extract its import (a warning as in the vervet case; a request, as in a courtship ritual; a threat, as with a canine growl), and decide on a response. Ostensive-inferential communication is much more flexible; indeed, it does not require a pre-established set of signals at all. Ron can indicate his desire for another drink by rolling his empty bottle of Coopers across the table. But flexibility comes at a cognitive price. As his audience, I have to wonder why Ron so overtly rolled the bottle in my direction, and construct on the fly a simulation of a chain of reasoning that would generate that act. Ron has to anticipate that I would notice, take his act to be a signal, and correctly infer—on the basis of a good deal of common knowledge—that beer was required. Misfires happen, but often communication runs so smoothly that no one is conscious of their inferential dance.

For reasons canvassed in section 1.4, we do not accept a sharp distinction between code-based and ostensive-inferential communication. But we agree that interpretation and its anticipation are central to contemporary conversational exchange, and we see our sender-receiver framework as naturally accommodating this fact. Indeed, as we said above, we think that interpretation and its anticipation were once even *more* critical to hominin communication. For regularized meaning is the result of a long history of successful, partially successful, and unsuccessful communicative attempts rather than their precondition. Once, much hominin communication would have been more like hopefully rolling the bottle than saying "It's time for another beer, and it's your turn to buy." So some ability to interpret, and to predict others' interpretations, was important early, not just now.

In the chapters that follow, we tie incremental advances in theory-of-mind capacities to incremental communicative advances, though often in a rather abstract way, reflecting the (current) lack of fine-grained evidence on how and when theory-of-mind capacities changed in our line. In our view, there are a few different and compatible ways theory of mind can become more powerful. Most obviously, the set of cognitive capacities an agent is capable of representing can increase. An agent who can represent the relations of seeing and knowing has a more powerful theory of mind than one who can represent the relation of seeing only. But there are other ways for theory-of-mind capacities to be upgraded even as the set of mental capacities one can represent is held fixed. It is plausible, for example, that one's theory-of-mind capacities increase as one's general conceptual repertoire expands in size. An agent with a larger conceptual repertoire is (presumably) able to attribute a wider range of mental states to others than one with a smaller repertoire. They may well be able to identify patterns in others' behavior that are not visible to the agent with the more limited conceptual repertoire, even supposing that the two share the same set of mental state concepts. If Thag has the concept of poison, and Thug does not, presumably Thag is better placed to notice and understand a pattern in Thagma's cautious exploration of novel fruits and nuts. If so, then theory-of-mind capacities might have increased in our line simply as a result of our ancestors growing their conceptual base. That, of course, is likely to be explained by a range of causal factors: intrinsic cognitive changes, but also changes in social learning, including how teachers modify the physical environments in which social learning takes place. One is unlikely to acquire the concept of glue until a glue is invented. Moreover, conceptual repertoire will obviously increase as communication becomes more sophisticated. Thag can then learn new concepts from Thagma.

In addition, theory-of-mind capacities may have increased simply as hominin working memory capacities increased (Planer 2021). Briefly, working memory is a set of domain-general cognitive capacities implicated in the selection, retention (minimally over a few seconds), rehearsal, and manipulation of mental representations. As Peter Carruthers (2015) persuasively argues, in essence these processes involve the controlled, top-down direction of attention to mental representations. Representations enter working memory when attention is focused on them. They remain there if and only as long as that focus remains, resisting the pull of various impulses and distractors. One critical process is rehearsal, which involves taking an action sequence offline and attending to the likely effects of performing that action (e.g., hurling a stone at some target). Generally speaking, manipulation of working memory content consists of an organized series of mentally rehearsed actions. On Carruthers's view, targeting some representation with attention causes it to be globally broadcast (Baars 1997; Baars 2005), at which point it becomes available for processing by a wide range of cognitive mechanisms, including ones that might not have otherwise had access to the representation. One possible upshot is an executive decision to attend to a new set of representations, making possible sustained, goal-oriented, reflective cognition.

Planer (2021) provides an account of how increases in working memory could have enabled our ancestors to make more flexible and more creative use of standing great ape theory-of-mind mechanisms; they could get more out of those mechanisms. Great apes have limited theory-of-mind abilities, but they can certainly recognize emotions and purposive behaviors in the here-and-now. So their theory-of-mind mechanisms are activated by representations of other agents: say a mental image of an agent oriented in a certain direction in some environment. By entering such a representation into working memory (either by recall, simulation, or sustaining an actual perceptual representation), that representation becomes globally broadcast and hence available for processing by much of the agent's suite of cognitive mechanisms. With an ability to rehearse, but with no upgrade in the mechanisms available to great apes, an agent can mentally test out various actions, attending to the predicted effects of each, and select a particular action in light of such predictions.[6] And going the other way, one can also engage in sustained, reflective reasoning about why some individual has acted in a certain way toward oneself.[7] In short, an enhanced ability to construct and entertain offline agent-representations, and to process them in sustained, cumulative ways, would make mindreading more powerful (in part through making it much more versatile). For it would enhance *any* of our cognitive skills: for example trackway reading, discussed at the end of section 2.4. In a word, then: One route to better mindreading is to invent better tools: for example, being able to represent the difference between what Thug believes and what he knows. Another is via an improvement in our mindreading *technique*, in how we used our *existing* mindreading tools, made possible by a more powerful working memory.[8]

Social learning. Even if syntax is genetically encoded and innate, language depends on social learning. For everything *specific* to a particular language must be learned. That includes its lexicon; the phonological and morphological regularities that constrain possible lexical forms in a language (there are words of Polish that are not possible phonological sequences of English); tense, aspect, and case systems, and the like. Humans specialize in social learning. Both Rob Boyd and Joseph Henrich have recently argued, very plausibly, that the most distinctive feature of our species, and the one that explains its numbers, geographical spread, and ecological impact, is our ability to accumulate knowledge across generations through social learning (Boyd 2016; Henrich 2016). Almost certainly, there has been selection on the mosaic of cognitive and social traits that make cumulative social learning possible. Amongst the products of that cumulative social learning are the languages we speak. However, while social learning extends far beyond language, competence in language and its predecessors has become essential to hominin social life; and the acquisition of this competence is an especially difficult social learning task.

To see this, contrast (say) acquiring a reasonably rich protolanguage with learning to craft a well-made handaxe. These are both impressive skills, with a certain similarity: they both involve the production and control of a long, partially preplanned sequence of precise elements. We think they both depend on social learning. Peter Hiscock has convincingly argued that stone-working skills are both difficult and dangerous to acquire,[9] but those costs can be markedly reduced by the active cooperation of a skilled model, at little cost to that model (Hiscock 2014). Even so, developing these skills takes sustained effort and practice. Protolanguage acquisition is even more difficult. (i) In the case of stonecraft, the novice has two sources of social information from the model. The novice sees the model's acts, or sequences of acts, and so has the opportunity to learn by imitation. But the novice can also see the consequences of the sequences of acts. He or she can see the stone as it is being worked, the *product* of the acts. Novices can learn by emulation from this product. They can see, for example, how the core is thinned over a series of strikes, or how large the flakes should be. Emulation is an information-rich channel, in part because you can feel the product, look at it from different angles, and so on. The product persists over time, too, so not all available information need be extracted at once. Chimp social learning of their foraging techniques is probably mostly emulation. In acquiring protolanguage, the second channel is closed. Neither gesture nor vocalizations leave an enduring structure to be examined. There is a process, but no enduring product. (ii) In the case of stonecraft, the novice can combine social with nonsocial, individual-exploration learning. The novice can experiment with stone, trying out different stone types, different hammers, different angles. Very likely there is some local variation in stone toolmaking techniques, but the workable options are sharply constrained by the world, by the properties of stone and its reaction to force. The techniques are not arbitrary, and so they can be discovered, in whole or in part, by individual learning. It would be no surprise to discover convergence in handaxe-making techniques in widely separated parts of the world, say India and East Africa[10] (though in fact common ancestry has probably played some role). It would be beyond surprising to find two communities that had independently built rich and largely identical protolanguages. In principle, a linguistic novice could experiment, trying out gestures or vocalizations at random and seeing what happens. But there are too many degrees of freedom in the choice of terms and how to organize them for that to be a realistic strategy, though novices can and probably do experiment with calls or gestures they perceive in use around them. So a novice's only access to the protolinguistic repertoire of her community is through social learning from that community. It follows that rich protolanguages can be built only when social organization and individual cognitive capacity combine to support reliable, high-fidelity, large-bandwidth social learning. That consequence will be important to the argument of this book, especially in chapter 7.

Michael Tomasello has identified a particularly important form of social learning, and one which may be unique to our lineage, the capacity to form—through experiencing or observing interactions—a "bird's-eye view" representation of a social interaction or collective action problem (see, for example, Tomasello 2009). A bird's-eye view representation is one that represents the structure of an interaction independently of the specific identity of the participants in the interaction. It represents *the role* of those participants. That is not always necessary: if a mob of noisy hominins bluff a leopard from her prey, no agent has to take any specific role. But if there is genuine teamwork, and especially if roles within the team vary, it matters. For example, a bird's-eye view representation of an ambush hunt would identify such roles as the drivers of the animals; lookouts; the ambushers in their specific locations. Bird's-eye view representation is very important in any form of teamwork in which agents might have to switch roles: holding a stone blank in one interaction; striking it in the next. It makes role reversal much easier. Humans, including quite young children, are good at switching roles in joint-action problems; great apes find it much more difficult. Since flexibility in collective-action problems is such an important aspect of human cooperation, bird's-eye view representation is of great general relevance to our social life. But it is directly important to language as well, as conversations are often instances of joint action in which an agent's role changes from occasion to occasion, and even moment to moment. Likewise, it must help, in taking advantage of another's innovation, if onlookers can represent themselves in that situation.

Prosociality. Language as we now have it depends on a social psychology very different from those of the great apes. We are much more socially tolerant and cooperative than the apes: a point Sarah Hrdy makes vividly at the beginning of her *Mothers and Others* as she imagines what would happen if one tried to fill a transatlantic jumbo with chimps rather than tolerant, resigned humans with their (comparatively) excellent impulse control (Hrdy 2009). Conversation—sharing information, planning, coordinating—is a special case of cooperation. It requires the same willingness to tolerate proximity for extended periods, and the same willingness to help, as other forms of cooperation. Michael Tomasello emphasizes the fact that unlike great apes, we often find joint action intrinsically, not just instrumentally, rewarding. But there are other motivational changes more specific to conversation. One is turn-taking. As we have already noted, Stephen Levinson has shown that the pace of conversation, with its rapid-fire shift between talking and listening, intensifies the processing demands on language use. But it also requires tuning nonlinguistic skills. As part of a conversation, agents must be sensitive to, and care about, the cues and signals that manage turn-taking, signs indicating (welcome) intervention points, confusion, loss of interest, impatience to change the subject, etc. Another is attention and intentional listening: there is a profound difference between having someone's conversation wash over you, as you daydream or focus on something else, inserting fillers at appropriate points[11] ("yes dear"; "is that right?"; "no, really?") and actually attending to your conversational partners. The evolution of distinctively human forms of social life depends (amongst much else) on evolving the ability to resist distraction, and that is true of our use of language, too.

Language is motivationally distinctive in a further way. We assume that the precursors of language originated as technologies of communication, to allow members of a community to coordinate; to make requests and issue instructions; to share information. Protolanguage was a useful instrument, and those uses selected for its preservation and elaboration. At some stage, participating in conversation became less instrumentally motivated. People now enjoy telling jokes, listening to stories, teasing one another, just chewing the fat. These forms of language very likely have, or have had, positive fitness benefits in defusing social tensions and increasing social cohesion and local identity (Dunbar 1996; Wiessner 2014). Listening to a story is enjoyable, and likewise conversation is often intrinsically rewarding. These social uses of language, or of protolanguage, signal an important change in our motivational psychology and very likely coevolved with those changes.

Uniquely human capacities? Short-term memory, executive control, semantic memory, mental models, theory of mind, social learning, tolerance, and prosocial motivation are all capacities that exist in other animals, though often in a much more limited or rudimentary form. Upgraded and suitably linked, they contribute profoundly to making language and its predecessors possible.

These capacities all play many roles in our cognitive lives. What about capacities that are unique to language? While we doubt that our mastery of language depends on cognitive capacities that are used only for language, we are certainly open to the possibility that language in part depends on abilities that are uniquely human. For example, the capacity to produce and understand signals having displaced reference might be such a capacity. With the possible exception of bee dance,[12] animal signals are about the here and now (though there is some suggestion that chimps can refer to an absent object by pointing to where it was; Bohn, Call, et al. 2015). The courtship display of a superb lyrebird is an invitation to mate directly, not a suggestion to meet behind the pub beer garden tomorrow night. Likewise, the vervet warning calls are indicators of immediate danger of specific kinds. We can use names (for instance) in the immediate presence of their bearers, as when we introduce a speaker. But often we use them when the bearer is absent, for example when gossiping behind someone's back. Yet even if this capacity is uniquely hominin, it is not used only in language, unless thinking is literally thinking in one's public language. We can think beyond the bounds of the immediate situation, and we can represent the elsewhere nonlinguistically, in maps and other representational formats. In any case, in the next chapter (section 3.3) we will offer an alternative view of the cognitive mechanisms on which displaced reference depends.

Perhaps the most famous candidate for a capacity unique to both language and humans is syntax. The immense expressive power of language derives in part from the remarkable combinatorial possibilities it allows, as in "The student who asked the embarrassing question at the last department seminar of the distinguished Polish professor . . .". It is certainly possible that this capacity is unique to humans and to language. There are other possibilities though, and in chapter 5 we tentatively develop one. We will argue that this capacity to organize and recombine items into a structured sequence, and the ability to use terms with displaced reference, depend on the evolution of complex technical skills, of the kind deployed in making handaxes. Those technical skills depend on the use of a mental template to plan and control action; a complex sequence is bought under internal control. Moreover, we shall argue that those sequences are structured and hierarchically organized. In our view, current neuroscientific and behavioral evidence suggests that syntax depends on capacities for recognizing and processing structure that we share with other animals (in particular, primates), though these are greatly enhanced in our lineage. Both displaced reference and syntax are the result of redeploying cognitive capacities that evolved to meet the special technical demands imposed by the Acheulian and later technological industries. Those capacities evolved initially for reasons other than their importance for communication.

2.3 Gesture and Its Importance

There is a natural explanation of the fact that great ape gesture shows more top-down control and more sensitivity to experience than does their vocalization. For the great apes are extractive foragers, often exploiting resources which vary on both local and regional scales, and which often can be harvested only with the aid of quite complex motor routines. To harvest such resources, agents must have visually guided control over their fine motor manipulation, and they need to be able to learn new routines, as their environment changes and as they shift through their local patch. The demands of their ecology require control and flexibility in their use of their hands (and the rest of their body). They are poised for the use of gesture, and given that great apes gesture at all, it is no surprise that they use gesture in learned and context-dependent ways. That is so even though the range of their gesture is very limited, for most gestures are requests. The great apes do not readily point informationally. Despite the limited range of great ape messages, the repertoires are quite large. It is claimed that gorillas have (at least) 102 distinct gestures (though these do not map onto 102 distinct requests). But though large, the repertoires are rather inflexible, because they are not readily expandable. For example, according to Genty and colleagues' analysis, gorilla gestures seem mostly to be species-typical items of behavior that have been brought under intentional, top-down control (Genty, Breuer, et al. 2009). Gorillas seem to have evolved the capacity to use their gestural options quite flexibly, but have limited ways of adding new gestures to their communicative toolkit. Their codes seem to have expanded only by associative learning, as directly functional action patterns became abbreviated and ritualized as signals. So while there is more capacity to add gesture than there is to add new calls, the system is by no means readily expandable.[13]

While there is not much flexibility over time, there seems to be a fair bit of flexibility at a time, for there does not seem to be any simple, one-to-one code of signal to response. Catherine Hobaiter and Richard Byrne have helpfully systematized reports on chimp gesture, and their database probably shows some gesture-to-response specificity. Gestures probably do not just mean: *pay attention and do something* (Hobaiter and Byrne 2014). But there are several gestural routes to each desired response,[14] and these are quite often given in series, when the communicative partner is initially unresponsive (thus satisfying one of the criteria for top-down control discussed in section 2.2). Moreover, the same type of gesture can be used to solicit different responses. In short, gesture and its use seem to show the reasonably advanced social intelligence that Seyfarth and Cheney take to be widely available amongst the primates, a kind of crude proto-pragmatics. The agent must select from a range of options, in ways sensitive to what their target audience can see and hear: "Gesturing of great apes is appropriately adjusted to the attentional state of the recipient. Silent, visual gestures are given mainly when recipients are looking; audible, visual gestures less so; and tactile (contact) gestures are given indiscriminately of the audience's attention" (Genty, Breuer, et al. 2009, p. 528).

Perhaps (though this is somewhat more contested) they are also somewhat sensitive to what others already know (Crockford, Wittig, et al. 2012). There is social intelligence on the receiving side as well, for gestures are often ambiguous, so the target of the gesture has to use contextual cues—some of which might be quite subtle—to identify their expected response. Others are less subtle: chimp males displaying their erect penis as a sexual request. Despite these easy cases, our best guess is that great ape receivers are quite good interpreters (again, congruent with Seyfarth and Cheney's claim that primates are well tuned to their social world). For there are plenty of reports of gestures being ignored, but there do not seem to be reports of frustration and conflict being caused by responding in an unintended way. There is even a modicum of evidence suggesting that great apes are sensitive to common ground: a recent experiment gave chimpanzees the choice between requesting a moderately desirable food, or requesting better food by pointing to a plate where such food had been. Would they make such points only to an audience who knew what had been on that plate? There was some evidence of that sensitivity (Bohn, Call, et al. 2016).

Thus in our view, baseline considerations about the great apes, and hence about the last common ancestor, make it more likely that communication expanded through an expansion in the number and precision of gestural signs. We are by no means the only defenders of this view: it is also found in Tomasello (2008) and Corballis (2009; 2011), for example. Moreover, if that were true, it makes a number of features of language and its evolution more readily explicable. One of these has already been mentioned: it leads to a much more natural explanation of the origins of structured communication. We take up this and related ideas in the next chapter, but two points are relevant here: coevolution between gesture and foraging skills, and iconicity. First, if hominin communicative capacities initially evolved as systems of gesture, the evolution of elaborated manual skill and the evolution of gestural communication would support one another. They would depend on the same fundamental cognitive, perceptual, and motor capacities. Both select for the capacity to learn, memorize, and fluently execute increasingly complex sequences. As technical skills became more complex and more difficult or dangerous to learn (it is very easy to cut yourself making stone tools inexpertly), social learning almost certainly became increasingly important. Here too capacities for toolmaking and for gesture-making overlapped. The social learning of technique, especially by imitation, makes the hand movements of others salient. Others will notice and respond to hand movements. What others are doing with their hands matters. That same focus of attention is needed for the social learning of gesture, especially as gestural sequences become longer, more elaborated, less error-tolerant. On the picture presented here, the expansion of gestural communication was facilitated by the evolution of technical skills far more elaborate than those found in great ape lives, and vice versa. We elaborate this picture over the next three chapters.

Second, iconicity: the sexual solicitation signal of male chimps is not arbitrary, and we see that as an extra consideration in favor of the view that gesture and mime were initially very important in hominin communication. In accounts of the distinctive features of human language, arbitrariness figures prominently. The idea is that there is no natural relationship between term and referent: 'dog' could just as well refer to cats. With few exceptions, this is true of spoken words. That is no accident: language affords the option of a natural correspondence between sound and object only through vocal imitation, and few referents make a unique sound that humans can easily mimic. This is much less true of sign; actions, for example, have characteristic movement patterns, but rarely have a distinctive sound. That opportunity is exploited even by the highly developed forms of sign in use by contemporary humans, humans who have all the cognitive machinery that has evolved to facilitate our use of language. Corballis remarks that, for example, perhaps 50% of Italian sign terms are iconic, and very likely many signs that are not now iconic derive from signs that were originally iconic (Corballis 2009, p. 32).[15] Many other signal systems—for example, many road sign systems—depend on conventionalized icons.

We can and do use arbitrary, purely conventional symbols, so we take it that this reliance on iconicity when it is readily available shows that even our modern minds are helped by iconicity, presumably because it is easier to remember or to recognize iconic signs. The expansion of communication in our lineage began with minds less well adapted to learning and using signs, and in social environments with much less support for the acquisition of these skills. Children are now born into an environment that is saturated with language, with repeated exposure to core items of vocabulary. With the emergence of Motherese, children's early experience seems to be saturated not just with language but with a special, infant-friendly version of language. As the role of communication in the social life of early hominins began to expand, their young would not have been so lucky. Stabilized, regularly exploited conventions of sign/object pairings were still in the future, so they would have needed all the help they could get. For agents without specific adaptations for symbol use, and who did not enjoy a developmental environment that scaffolded sign acquisition, iconic signals would offer important advantages. Gesture would have offered those advantages much more freely than sound.

Not everyone is convinced that the potential iconicity of the gesture-target relationship supports the idea of a gesture-led expansion of early Pleistocene communicative capacities. In particular, Liz Irvine has argued that developmental evidence (and perhaps comparative data from chimps) suggests that iconicity does not help, and may indeed make the project of coining and understanding representations with displaced reference more difficult (Irvine 2016). She offers four main considerations in favor of this somewhat counterintuitive conclusion, though they are mostly variations on a theme: developmental evidence suggests that infants and toddlers pick up arbitrary symbol-referent relations faster or more fluently than they do icon-referent relations.

First, she points out that children are only able to use resemblance to infer meaning at about age two, and it is not really stable until about three, whereas they can map words (and other signs) onto their targets from about 18 months (p. 228). Second, infants in the age ranges of two to four do not interpret new gestures in the way they standardly interpret new vocalized words: assuming that a new symbol must name a novel object (or action). They do not interpret new gestures this way, even when the gestures are given in the same kind of context that triggers the novel symbol/novel meaning strategy with words (pp. 130–131). Third, she reviews evidence that it is only at about three that children can use models and other depictions as information sources about their targets. Moreover, the more interesting a depictive object is in its own right, the more difficult it is for young children to use that object as a representation. They find the "dual representation" this would require cognitively challenging. Finally, she suggests that iconic representations are suboptimal for actions controlled by system 2 cognitive processes, because iconic representations tend to trigger automatized responses. The research here originally derived from famously amusing experiments by Sally Boysen on chimps. Those chimps found it almost impossible to learn to point at a small pile of food in order to get a larger one, but could learn to do this when the actual food piles were replaced by arbitrary symbols. Notice, though, that while the shift to arbitrary signs helped the chimps communicate their desires, they were not struggling with iconic signs but with counter-iconicity: they had to point to a small pile in order to indicate a desire for a large one. It is as if they had to produce an icon of a mouse to refer to an elephant, and no one suggests that counter-iconicity helped our ancestors. Somewhat similar findings were replicated, explored, and extended with young children. Irvine suggests that this tendency to trigger associative response makes icons ill-adapted to represent beyond the here and now.

While our case for a gesture-led expansion of communication does not rest heavily on our claims for iconicity, we are not convinced by Irvine's argument. Taking the easiest case first: in normally vocalizing households, by the time toddlers are in the two- to four-year-old range, they have had thousands of hours of exposure to vocalized language, and have had ample opportunity to learn that lexical items are structures of phonemes, not gestures. In view of that experience, the fact that they have different interpretive strategies for gestures and vocalized words is no surprise. Second: one way to see this evidence is that *sapiens* infants can use icons extraordinarily early: somewhere between two and three. Moreover, the fact that infants love and understand soft toy animals suggests that some forms of dual representation come on stream remarkably early: Sterelny's two-year-old knew her small toy panther was a panther, but a toy panther. We think it is astonishing that by two, a child can begin to use icon-target similarity to infer from icon to target. If there is a puzzling datum here it is that 18-month-olds supposedly understand symbols as symbols. If we were to accept that datum at face value, it suggests to us the presence of a late-evolving adaptation for word learning, of the kind embraced by the "natural pedagogy" theorists (Gergely and Csibra 2006; Gergely, Egyed, et al. 2007). But given the severe limits on an 18-month-old's capacities to act, it is very unclear that (say) the word "Peter" in their mouths is a symbol in roughly the same sense that it is in our mouths.

Finally, let's take up the idea that by triggering automatized associations, iconic representations can short-circuit executive function, interfering with deliberative response. There is no doubt this can be true. Much representation in both pornography and advertising is designed to short-circuit such executive function, though notice that advertising and pornography typically involve hybrid representations, with a mix of icon and symbol. Symbols too can turn on automated responses. The skull and crossbones label for poison, and red as a danger signal, are not icons. In any case, we do not see the evidence for claiming that this is a general feature of iconic representation, or that it poses a special problem for understanding icons as having displaced reference. Thus Irvine writes: "action-based gestures and pointing, along with elicited responses, can take you fairly far in communicative contexts set in the here and now. The difficulty arises in situations where one needs to refer to things out of sight, or in the past or future; here one needs symbols. And this is where the idea of being able to inhibit routine reactions becomes particularly important" (Irvine 2016, p. 235). We do not see this. If gestures or mimes routinely trigger automatized responses, that is as big a problem in the here and now as it is in the case of reference beyond the here and now. We do not communicate about the here and now just to trigger a stereotyped response. Irvine's elaboration of this argument seems to presume this design link between gestural communication and stereotyped response. Thus she goes on to suggest that the function of teaching skills through gesture or mime is to elicit "reactive behaviours" (p. 236). Again, we do not see this: teaching skills through gesture and mime is important precisely when the novices do not have automatized routines to trigger. The mimes are props to help them develop those routines. In short, Irvine is right to point out that there is a class of icons that are likely to trigger an automatized response, subverting executive control. But this is not a general feature of icons. It is not true that icons are interpretable only because and as they plug into and drive associative responses. Crocodile icons on North Queensland beaches have been known to trigger games of chicken, and attempts to lure crocodiles in, as well as due caution at the water's edge.

We cannot use developmental data about young modern humans as evidence about the cognitive capacities of adult and near-adult Pleistocene hominins as they began to construct the foundations of what would become language (and nor does Irvine). But the fact that very young children can use iconicity as a guide to interpretation suggests that it is not especially challenging; it is some support for the view that quite encephalized middle Pleistocene hominins could use iconic signals. Even so, let us emphasize again: in our view, iconicity helped, perhaps especially in coining or learning new signs. But it was certainly not essential. As Antonella Tramacere and Richard Moore point out, a gesture-first model of language evolution can do without a passage through iconicity, using pointing as a scaffold to introduce arbitrary signs (Tramacere and Moore 2017). We remain skeptical of the complete rejection of any role for iconicity. But we agree that pointing was important early on, and some other early gestures may not have been iconic. The same was probably true of calls, if and to the extent that the early expansions of communication involved vocalization.

Finally, does Kanzi, the celebrated language-using bonobo, undermine the view that iconicity scaffolded early protolanguage? Not in our view. We will discuss Kanzi in relation to syntax in section 5.2. But he is relevant here too. Kanzi and his later companions mastered a significant arbitrary vocabulary, without the extensive and ecologically implausible training regimes of somewhat earlier great ape language experiments. However, it is important not to overinterpret these enculturated bonobos (and the even more surprising case of a dog with a large receptive vocabulary).[16] While Kanzi and Panbanisha (Kanzi's half-sister) were not explicitly taught, they *were* raised in a language-rich environment. So while we know they had the cognitive capacities to take advantage of a sign system with an extensive repertoire of arbitrary signs, we do not know they had the cognitive capacities to grow such a system.

2.4 Early Hominins: The Cognitive Consequences of Bipedalism

We have placed the split between the hominin and *Pan* lineages at about 7 mya. Molecular clock methods of estimating the divergence date based on the rate of DNA mutation yield a range of possibilities, but we think 7 mya is unlikely to be a large undershoot or overshoot. Between that divergence point and something like 2 mya, the physical record is obscure, challenging, and controversial. The first fossils supposedly representing the appearance of our genus *Homo* are generally assigned to *H. habilis*, and these are somewhat older than 2 mya, though they all fall in the Pleistocene. The first known stone tools, at 3.3 mya, are considerably older, and there is significant uncertainty about the identity of the hominins who made them. The standard genealogy of hominin evolution supposes that the australopithecine ancestors of our genus were a lineage of hominins with approximately chimp-sized brains (relative to body weight). That lineage was quite bushy, featuring a considerable number of australopithecine species, sometimes with several living at the same time. Some were quite robust, especially in the face and jaws, and these probably increasingly specialized in tough, difficult-to-chew plant foods. Other members of that lineage were more gracile, more generalized in their diet, and over time became more bipedal (as perhaps were the robust australopithecines) and less sexually dimorphic. This gracile lineage of the australopithecines is the presumptive direct ancestor of the *Homo* genus, and the presumptive maker of the Pliocene stone tools. Very likely these were used for many purposes, but one was to cut meat off large herbivore bones.

This standard picture has a number of serious problems. To begin with the simplest point, the idea that there is a first "true human," and that the genus *Homo* is a lineage of all and only the true humans, is biologically implausible. There is a powerful case to be made for the objective reality of biological species (though identifying them is another matter), but there is no case for the objective reality of higher taxonomic ranks: genus, family, and so on (Ridley 1986). The lineage of a species and all and only its descendant species is a real segment of the tree of life. But which segment of the tree we choose to call a genus, a family, and so on is a matter of convention and convenience. Even were we omniscient about the genealogy of the hominin lineage, there would be no identifying the first true species of humans. A more serious point regards the actual identification of extinct hominin species. Paleoanthropology texts are littered with an ever-increasing number of named species, often based on one or a few specimens. Almost all of these are fragmentary and incomplete, some very radically so. That is especially true of the Miocene and Pliocene hominins.[17] The biological meaning of these named taxa is very unclear indeed. Many aspects of the living phenotype of the species from which the fossils derive are unknown. The natural range of variation at and across time is even more hidden. Consider the variation in existing human body shapes and sizes. Were future paleoanthropologists to find even full fossils of a Maori second row forward and a gracile Hadza forager (about 60% of the height and perhaps 40% of the body weight of the rugby-playing Maori), it is unlikely that they would be recognized as members of the same species. It goes the other way as well. There are many contemporary examples of species whose distinctness was recognized initially only through DNA sequencing: cryptic speciation has turned out to be quite common.

The fossil record tells us that over 7 million years, hominins have changed dramatically, and it tells us that over much of this period there was a good deal of variation present at most times. It tells us much less about how that variation was parceled up into species, reproductively isolated at a time and on independent evolutionary trajectories over time. Identifying specific ancestor-descendant relations between fossil taxa is even more problematic. That is especially true if over much of hominin evolution there were indeed several closely related species in existence at any one time. Inevitably these sibling species would leave very similar fossil traces, and equally inevitably, the accidents of preservation could delete many such species from the record. It is thus quite likely that while we may think that *A. sterelnyi* was the direct ancestor of *H. planerii*, the real ancestor is *sterelnyi*'s undiscovered sibling species, *A. godfreysmithii*. In the chapters to come, we will write of australopithecines, habilines, erectines, heidelbergensians (and for the uninitiated we give brief descriptions of each in our glossary). But in doing so, we have in mind not specific taxa but grades: that is, clusters of morphological, behavioral, and cognitive features characteristic (respectively) of hominins in the late Pliocene, the early Pleistocene, the mid Pleistocene, and the final third of the Pleistocene. We take no stand on the extent to which each of these grades corresponds to a single taxon, several taxa, or no taxon at all (if, say, a cohesive erectine lineage gradually evolved the characteristics of the later heidelbergensian grade).

Given these many complications, we shall not attempt to reconstruct an incremental narrative of morphological, behavioral, and cognitive change in late Miocene and Pliocene hominins. The data are too fragmentary, and the ancestor-descendant links are too conjectural. Instead, we will focus on the state of play in hominin lifeways in the late Pliocene, and around the Pliocene/Pleistocene boundary, to get a rough estimate of some of the behavioral, social, and cognitive capacities of those hominins generally supposed to be the immediate predecessors of the so-called *Homo* species. As we see it, the key fact is that toward the end of the Pliocene (at the latest) there were fully bipedal hominins, and that the technology and foraging ecology of those hominins had diverged from great ape norms. The specific identity of that species, or those species, does not matter to our argument, so long as one of them was ancestral to Pleistocene hominins. We reserve our discussion of the cognitive and communicative implications of technology and foraging for later chapters. Here we take up the cognitive implications of bipedalism itself. For we shall suggest that obligatory bipedal lifeways in themselves select for cognitive changes that helped make the richer technological and social worlds of the Pleistocene possible.

There was once a standard narrative of the origin and spread of bipedalism. East Africa was the epicenter of hominin evolution, and it lost its continuous forest cover as a result of some combination of regional factors (rifting and uplift causing rain shadows and volcanism) and global factors with the shift to the drier and less stable world of the Pleistocene. There was a shift to an environment dominated by open woodland and savannah, perhaps with remnant forest patches in favored areas. Bipedalism was an adaptation to this more open environment, with its patchier, more scattered resource profile. This profile forced longer daily and perhaps seasonal journeys on these hominins. Continuous forest is more homogeneous, with resources spread more evenly. So these more open-country hominins faced an environment with less cover, requiring more movement, and with crucial resources more concentrated in scattered patches.

This picture is probably too simple. For one thing, *Ardipithecus ramidus* fossils (4.4 mya, Afar, Ethiopia) seem to belong to a forest-dwelling bipedal hominin, and some australopithecines seem to be adapted both for climbing and for bipedal travel, presumably between trees or tree patches. Great apes, forest animals, are capable of bipedal motion and stance, though for brief periods and over short ranges. So it is conceivable that there is a route to bipedalism within a lineage that is still essentially dependent on trees: modern humans can still climb trees well, though they cannot sleep safely in them. Another possibility, vigorously urged by Kim Shaw-Williams, is that bipedalism is an adaptation to wetland, riverine, and coastal life (Shaw-Williams 2017), a possibility supported by the recent find of Miocene footprints on tidal muds near Crete.[18] This need not be a fatal conflict: as Maslin rightly points out, bipedalism may well have evolved more than once, and under different selective regimes (Maslin 2017, pp. 75–76). Nonetheless, whatever its origins, by the late Pliocene to early Pleistocene, East African landscapes were drier, more open, more seasonal than they had been a few million years earlier, and those landscapes were inhabited by hominins who were not just episodically bipedal. They were obligate bipeds, or very nearly so. What were the cognitive consequences of this change?

Ben Jeffares points out that in earlier work on human evolution, the evolution of bipedalism was seen as central to our distinctive nature (Jeffares 2014). He thinks, and we agree, that the switch of scientific attention to larger-brained Pleistocene hominins and to the relative importance of social and technical intelligence has led to an underappreciation of the importance of bipedalism. One suggestion in this regard is simple and plausible. Once hands were free from the stresses involved in knuckle walking, a constraint was lifted from hominin evolution: hands can become adapted to a greater range of manipulative tasks, involving both precision and power grips. Hominin shoulders and arms can become adapted for throwing: for the use of projectiles, even as simple as rocks, in defense, in bully scavenging, and eventually in hunting (Ambrose 2001). Jeffares himself focuses on the implications for memory, navigation, and the use of space. One of bipedalism's advantages is that it is more efficient than knuckle walking, and it needed to be as East African environments become patchier and more seasonal. In such circumstances, hominins would have had to travel further for food, water, and shelter. They would need larger home ranges; indeed, ethnographically known African foragers have much larger range sizes than those of chimps (Layton, O'Hara, and Bilsborough 2012). That in turn increases demands on navigation and memory: those hominins needed to remember watering holes and safe shelters, and how to get there. Moreover, Jeffares points out that bipedalism opens a new way of using space. Bipedal hominins can carry loads, so they need not process and consume resources at the place of harvest. Let us suppose that the Pliocene evidence of access to meat from large animals (presumably at least initially by scavenging; see Thompson, Carvalho, et al. 2019) records reasonably regular events (McPherron, Alemseged, et al. 2010). If so, there would be significant selection pressure to take advantage of that opportunity. But selection would also favor getting away from carnivore killing fields as quickly as possible, especially before nightfall. Pliocene hominins were not physically imposing. In turn this implies selection for some increase in planning depth and impulse control. Chimps eat what they find on the spot. Moreover, if Pliocene hominin bands were able to roughly disarticulate a carcass (or the remains of one) and carry the chunks to somewhere safer, they also needed a fair degree of social tolerance, to avoid the band disintegrating into chaotic squabbles about who had what.

We do not think this suggestion of social tolerance in the late Pliocene is fanciful, for we think habitual bipedalism, and especially obligatory bipedalism in more open environments, selects for a more cooperative social environment. Bipedalism is a challenge to mothers of babies and infants. For their infants cannot ride clinging to their backs as do great ape infants. A bipedal mother seems to have four options: some version of creching (leaving her infant behind with someone, implying reproductive cooperation); carrying an infant in a sling or basket (which implies soft-material technologies much superior to anything in the great ape world); carrying the infant herself; or carrying it with the help of others (again, implying cooperation). Solo carrying is a tough ask. Evidence from both Hadza and San foragers indicates that mothers foraging with infants have their foraging efficiency much reduced, despite having their infants in a sling (Lee 1979; Marlowe 2010). Without a sling, the energetic demands on a mother carrying a baby or infant in her arms over considerable distances would be significant, especially as she already has the considerable burden of manufacturing breast milk. That is true even if a young infant could ride on her shoulders—far from easy over rough or steep ground, and imposing a considerable energy burden if the mother foraged over considerable distances. Of course, Pliocene infants would not be helpless nearly as long as *sapiens* infants. But even once they are walking, could they walk as fast or far as the adult party? Moreover, in more open habitats they would be very vulnerable (Hart and Sussman 2005). Younger forager children generally play and forage around the campsite, normally with a few adults in attendance. They do not typically accompany adults on foraging expeditions.[19] In short: we find it hard to see how bipedalism of the kind that involves significant daily travel over a substantial home range could be viable without some form of reproductive cooperation (though perhaps sling technology would be an alternative possibility).

Likewise, we think that predation threat-selected for cooperation, or at least for better social tolerance. We have in mind the threat of nocturnal predators. Chimps nest relatively safely in trees. Moreover, they, and even more so gorillas, are much more physically imposing than (especially) gracile Pliocene hominins. What would obligate bipeds do to find safety at night?

Clive Finlayson suggested that rocky outcrops near water were the most likely refuges (Finlayson 2014). We think that is plausible, but only when combined with resting in close proximity for mutual support. A band of australopithecines scattered here and there through a rocky outcrop would be very vulnerable to leopards and any other reasonably agile nocturnal predator. But safety in numbers requires at least tolerance of long periods of close proximity, and probably a willingness to provide more active support. So putting this together, a shift to bipedalism was probably accompanied by enhanced memory and navigation, some increase in planning depth and impulse control, greater social tolerance, and perhaps tolerance with active cooperation.

We will end this section with a more speculative idea, though one that we find very plausible. Bipedalism alters the relative role of smell and vision. As a hominin stands on two feet, his or her head is lifted further from the ground, further from the odor-bearing substrate. So the information flow from smell diminishes. At the same time, the higher location of the eyes improves an agent's field of view. The value of vision goes up. Kim Shaw-Williams links this change in the default information flow reaching the hominin mind to his account of the importance of wetland foraging in the evolution of bipedalism to explain a distinctive hominin skill (Shaw-Williams 2014; Shaw-Williams 2017).[20] Apparently, we are the only animals that systematically exploit visual trackways: footprints and other marks of passage. This skill is fundamental to many forms of foraging (Liebenberg 1990; Liebenberg 2013). But it is also important in social interaction, for foragers recognize the footprints of all the members of their social circle (a fact that makes illicit assignations more difficult). It is also important in navigation. Marking your own trail makes it possible to retrace your route, and Shaw-Williams points out that navigation is challenging in many wetland environments, as tall grasses greatly restrict an agent's field of view and mudflats and sandbars change after every rain.

We know these skills are highly developed and very important in historically known foragers (Liebenberg 1990; Liebenberg 2013). Obviously, it is very difficult to date their first emergence, but Shaw-Williams points out that at least a minimal form of trackway awareness is manifest by the Laetoli footprints, a trail of hominin footprints preserved in ash from over 3.5 mya. For those record one hominin stepping into prints made by another. To step in another's footprints so carefully, those prints have to be salient to you. We do not know whether the trailing hominin was immediately behind the two in front, or was following sometime later, which would require still greater awareness of their significance. Trackways are cues rather than signals[21] (though they can very readily morph into signals, as when you make sure that traces are there for you future self). But they are a very unusual kind of cue. Their directionality means they carry information about the direction of travel, and hence about where the track maker is now (or was, somewhat earlier). Tracks and their makers are also displaced in time, and often in ways detectable by the trackway reader: tracks can be very fresh, somewhat recent, or clearly days or months old (tracks in clay baked hard, for example). These differences make a difference to the significance of a track; trackway readers need to note the tracks' age. Reading tracks trains the mind in reading traces about the elsewhere and elsewhen. We suspect that Shaw-Williams is right, and late Pliocene hominins (or early Pleistocene hominins) were trackway readers, though of a fairly minimal kind. If so, to some degree they were primed to treat the actions of other agents as information sources about events beyond the here and now.

Clearly, there is much about the lifeways of late Pliocene hominins that we do not know. But there is evidence of considerable divergence from lifeways typical of great apes: in their use of space; in the use of stone tools; in their diet, with a greater exploitation of animal products; in greater cooperation and social tolerance. These factors suggest that their communicative capacities and needs were beginning to diverge from great ape patterns too, especially if their cooperation involved any forms of teamwork and collective action. We will not attempt to guess those communicative capacities, for our knowledge of their social and economic life remains too patchy. Instead, in the next chapter we develop this picture further, with two main aims. One is to paint a picture of the lifeways of the erectines. These hominins emerged at about 1.9 mya, and we accept the widely held view that they were much more similar to recent humans than were erectines' immediate ancestors. Our aim in reconstructing their lifeways is to establish from the archaeological evidence that (i) they had cognitive capacities that would suffice for a reasonably sophisticated protolanguage; and (ii) their lives would benefit from such a protolanguage. We infer that they probably had such a protolanguage. We then develop a case for thinking that their protolanguage had three critical language-like elements: (i) it was structured; (ii) it included displaced reference; (iii) new items could be added with some facility.

A final remark. In the reconstruction that follows, we will primarily focus on the interactions between foraging (and more generally subsistence), social networks, cooperation, and communication. However, these interactions play out against a background of an increasingly dynamic and variable environment. The Pleistocene (especially) is characterized by a cycle of colder (and often more arid) periods interleaved with warmer ones, with that pattern becoming more intense and more rapid in the second half of the Pleistocene, with the climate becoming more variable on a range of temporal scales. Our epoch, the Holocene, is recognized as distinct from the Pleistocene because it marks the end of that extreme variability (or perhaps a pause in it). These environmental trends shaped hominin evolution. They most directly impacted subsistence, but also social networks, cooperation, and communication. These in turn impacted the selective environment: most obviously, the social aspects of that environment, but human action can and does have directional and cumulative impacts on the biological and physical environment, for example by using fire as a hunting tool. Hominin evolution in the Pleistocene, with its expansion in neural capacity and increasing dependence on cooperation, information, and technology, was in part a response to external forcing. But feedback between social organization, cooperation, and communication was important too, as was great regional variation in hominin environments as the Pleistocene progressed (Gamble 2013). So the selective forces acting within hominin populations changed over time. However, these coevolutionary interactions took place against an increasingly challenging background.[22] This issue becomes prominent again in section 7.3, where we discuss potential external drivers of the very late Pleistocene changes in human cooperation.

Peter Godfrey-Smith discusses the contrast between externalist and internalist models of evolutionary change. Externalists weigh most heavily the effects on lineages of independently changing environmental changes; evolutionary response to changes in climate or sea levels are paradigms of the phenomena externalists regard as typical. Internalists (obviously) emphasize internal, coevolutionary factors: changes in males driven by female choice is a paradigm (Godfrey-Smith 1996). The distinction is important but not exhaustive: for our part we focus on internal interaction, but against a background of major environmental change.[23] As we see it, the environment is important, but not on its own determinative. For example, marine isotope stage 5 (between approximately 130 and 80 kya) was a comparatively warm wet period (and Marine Isotope Stage 5e, 124–119 kya, was probably warmer than our current climate) but there is no sign of experiments in agriculture during that interglacial. The climate change of the Holocene, with its more benign and less variable weather, was probably necessary for agriculture and sedentary society to begin, but it was not sufficient: it required cultural and technical co-conditions. We take this example to illustrate the more general pattern of interaction and co-causation between factors internal to evolving hominin lineages and environmental factors changing independently of those lineages.

3 Let's Sign (and Speak) Erectine

3.1 The First Hominin Success Story

If our speculations toward the end of the last chapter are correct, then pre-*Homo* late Pliocene hominins had already diverged from the other apes in their cognitive and social sophistication. However, on balance they were still more apelike than humanlike. That changes with the rise of *Homo erectus*. In the eyes of many, the evolution of the erectines, at about 1.9 mya, crossed a divide: we had for the first time a species strikingly similar to anatomically modern humans (AMHs). Indeed, Clive Finlayson goes so far as to treat *erectus* as a subspecies of *Homo sapiens, H. sapiens erectus* (Finlayson 2014). From the neck down, erectine bodies were similar to the bodies of contemporary humans, though somewhat smaller than the high end of the contemporary range: they seem to have varied from about 160 to 180 cm (5.3 to 5.8 feet) in height. Their brains were significantly smaller than those of contemporary humans, but the erectines were very significantly encephalized in comparison to earlier hominins, averaging about 950 cubic centimeters of brain volume (Maslin 2017, p. 32). Human evolution is not just the evolution of supersize primate brains; but in our view, with the arrival of the erectines we have the first compelling evidence of a way of life that depended on cognitive, social, and communicative capacities far in advance of those known from great apes. In our view, the erectines needed, and had, a basic protolanguage. Initially, this was probably a small set of signs used with some fluency to coordinate core subsistence activities, gradually expanding as economic and social life became more complex, and as communication became less tied to gesture and to iconicity.

We base this claim on converging lines of evidence from five sources: (i) evidence of erectine large- and medium-game hunting; (ii) diet; (iii) changes in life history patterns toward those of more recent hominins; (iv) the cognitive and social demands of new technology; and (v) geographic distribution. There is a common pattern in the discussion that follows: we will show that erectine lifeways depended on having cognitive and social capacities that would suffice for a simple protolanguage, including capacities for individual innovation and social learning; and we will show that in those lifeways, there would have been significant selection for something like a protolanguage. Thus they probably had one. They were not tied to the present. They could anticipate future wants and act appropriately. So they could think causally, not just about interactions in front of their eyes, but about what was likely to happen, given one choice rather than another. They led more cooperative lives, which in turn required further theory-of-mind capacities. These factors in combination suggest that erectines lived a more planned and organized, a more information-dependent, and a more cooperative life than earlier hominins. Moreover, the forms of cooperation on which they depended required coordination and hence communication. Other social carnivores (wolves, African wild dogs, hyenas) cooperate in taking challenging prey, but they are much more tied to specific resources and a specialized (and perhaps genetically canalized) hunting pattern. In the rest of this section we fill out these ideas about erectine lifeways, and in the rest of the chapter we explore their implications for communication and protolanguage.

Hunting. We begin with hunting, for this is the only controversial claim of the five we will make about erectines and their lifeways, and it is one that is consequential for the view we will develop about their cognition and communication. The controversy dates back to (at least) the 1970s, with Glynn Isaac developing a view of erectine hunting that linked it to sharing, to central-place foraging, to male cooperation, and to a male-female division of labor. Isaac's picture was vigorously challenged by Lewis Binford, who argued instead that erectines were obligate but marginalized scavengers, getting very late access to carcasses (see Lupo 2012 for a brief history of these earlier debates). Late-access carcasses were a valuable but largely uncontested resource for hominin scroungers because they were able to get marrow and brain by using rocks as hammers to smash open large leg bones and brain cases (a picture that makes the investment in Acheulian technology rather mysterious, as cobbles and even unmodified stones can be bought into action as crude but effective stone hammers). In the last fifteen years or more, the case for erectine medium-game hunting (and somewhat later, larger-game hunting) has been made most consistently by Henry Bunn and Travis Rayne Pickering, and our view here has been shaped by their arguments (Bunn and Pickering 2010; Pickering and Bunn 2012; Pickering 2013; Domínguez-Rodrigo and Pickering 2017).[1]

They begin with the point that erectine hunting would be no surprise, as hunting by the *Pan* lineage strongly suggests that the last common ancestor of the *Pan*/hominin lineage hunted, a point reinforced by the observation that chimps tend to hunt more in the more seasonal parts of their overall range, together with the fact that the environments of Pliocene hominins were at least as seasonal as the most seasonal of current chimp environments. Moreover, they point out that savannah chimps use very simple wooden spears to kill bush babies safely, at a distance, and there is a very smooth and incremental path of technological improvement that takes a spear-armed hominin from the simple sticks used by chimps to high-velocity, tipped, woomera-launched javelins of the late Pleistocene and Holocene. On their view, large- and medium-game hunting developed from a platform of Pliocene small-game hunting plus occasional opportunistic scavenging (Pickering and Domínguez-Rodrigo 2012). As they see it, without stone tools, passive scavenging would not deliver meat regularly enough to make much impact on seasonal resource shortages, as the resources (regularly) available by scavenging required stone tools to break open large bones for marrow, and the skulls for brains. Likewise, the thick skins of large animals could not be cut by Pliocene hominin teeth; they needed an edge. Early hominins lacked such tools until perhaps 3.4 mya (and those very early dates are controversial).[2] Pickering and colleagues conclude that since isotope evidence suggests that early hominins had access to at least some meat, they hunted small animals. In this picture, the Pliocene-Pleistocene transition is a transition from small-game hunting and perhaps some passive scavenging to more active and probably aggressive scavenging, made profitable by stone tools which made it possible to harvest meat, marrow, and brains, together with some hunting, probably involving wooden spears.

Why think that *erectus* hunted large animals? There is archaeological evidence that African erectines had access to their carcasses. They exploited the meat and/or bones of antelopes and animals of similar sizes. Moreover, hunting was physically and technically feasible for erectines. There is experimental evidence that sharpened but untipped wooden spears can penetrate hide (zebra hide) if thrown from close range (Pickering and Bunn 2012, p. 161), and that they can hit with some accuracy and force up to about 20 meters (Milks, Parker, et al. 2019). Furthermore, erectines were morphologically equipped to throw such spears (Roach and Richmond 2015). But did they hunt or did they scavenge? Much of the debate about early erectine hunting turns on the interpretation of bone assemblages found in conjunction with lithics at various East African sites dated between about 1.8 and 1.6 mya. Especially critical is FLK Zinj, Bed 1, Olduvai, dated to about 1.84 mya, but two other equally early Bed 1 sites are also important: the Philip Tobias Korongo (PTK) site and David's Site (DS); both discovered since 2010, within 500 meters of FLK Zinj (Pickering and Bunn 2012; Domínguez-Rodrigo and Pickering 2017). These sites seem to show early access, perhaps even first access, to the carcasses from which the bones derive. For the FLK Zinj site has produced many long limb bones that would have been rich with meat and marrow, and these have defleshing scars. That is not the assemblage or use pattern we would expect if hominins had late access and harvested only marrow. We would then expect shattered bone and no defleshing scars. Did hominins have first access to these carcasses; if so, were they the killers? There has been a long and still unsettled exchange between Bunn and Pickering on the one hand and Blumenschine (with various coauthors) on the other as to whether these bones are marked by carnivore teeth. If they are marked by carnivore teeth, that might suggest power scavenging had played an important role in the formation of these assemblages of bones.[3] Even if that were true, if the bones were also significantly defleshed by stone tools, that is evidence of early hominin access.[4] Indeed, it has recently been argued that the transition from the lightweight Oldowan sharp-flake technology to the much more robust Acheulian handaxe technology was driven by an increase in access to large animal bodies, and the need for better butchery tools to process these carcasses (Toth and Schlick 2019).

Most importantly, though, the bones seem to come from animals of the wrong age if the hominin bone assemblage was mostly the result of hominins pirating carnivore kills (Bunn and Pickering 2010). For in that case, we would expect mostly young animals or old ones; these are the main targets of predators. That is not what we find (Bunn and Pickering 2010). Most of the bones are from adult males. This evidence suggests not just hunting but *directed hunting*. Let's suppose then that erectines were hunting reasonably regularly. How did these early erectines hunt? We shall explore some possibilities, but our crucial claim, and the reason why hunting is important to us, is that all the credible possibilities depend on far more extensive cooperation, coordination, communication, and social learning than is seen in the great apes.

In our view, the most plausible model of erectine hunting is ambush hunting. Hunting by stalking (or pursuit) of adventitiously encountered game is most unlikely to be regularly successful. For it would surely take multiple impacts from untipped spears to cause reasonably immediate crippling or killing damage. Successfully hunting medium-size bovids with thrown wooden spears would take weight of fire, and that in turn would make the problem of concealment for a group trying to get close enough very difficult. Bow-armed hunters that hunt by stalking do so in very small parties (often of one), even though they do not have to get as close as spear-throwers. Multiple bodies are a major liability.

In contrast to stalking, ambush hunting from cover is reasonably safe and makes possible a broadside weight of fire (so long as there is sufficient cover for a party). That fits the bone assemblage data too: hunting from ambush makes target selectivity possible, picking the most valuable target from a group, rather than the closest or most vulnerable, as in a stalk or a pursuit. Ambush hunting requires information and coordination, especially if the ambush involves driving the prey past the ambush point. For in contrast to animals who ambush hunt, humans lack specialized biological weapons and hunt animals as large as, or larger than, themselves. Indeed, if Binford is right, ambush hunting with short-range weapons needs a lot of information (see Binford 2007, pp. 196–202). Binford points out that hunting with weapons of the kind Bunn and Pickering have in mind can indeed be quite successful. But success with such simple weapons is almost always associated with catching the prey at some disadvantage: by catching them in physically vulnerable circumstances, or by exploiting some known, specific vulnerability in their escape behavior.[5] If Binford is right, regular success at ambush hunting depends on the hunters having rich, nuanced natural history information. That in turn presupposes quite extensive social learning, perhaps even explicit teaching. If erectines were reasonably successful ambush hunters at 1.8 mya, they were able to communicate well enough to coordinate in hunting and, probably, to facilitate social learning. Such hunting also requires impulse control and great social tolerance: those in concealment have to sit quietly, in very close proximity, even though ants, flies, and various other insects are likely to be feasting on them. Richard Gould vividly describes the discomforts of ambush hunting for emu, and how stolid the hunters were (Gould 1969).

There are other possibilities, but crucially they too imply quite impressive forms of coordination and social learning. Pickering himself suggested that erectines hunted by arboreal attack (Pickering 2013). The idea is that early hominins perched concealed in trees and (repeatedly) drove wooden spears into the spinal cords of large animals passing underneath. As Pickering notes, this method solves the problem of concealment. Perhaps with the exception of hominins up trees, the predation threat to large animals is from the ground, not the air, so they have not evolved to look up (whether they could learn to do so is a more open question). He also presents some ethnographic data in support of the suggestion. Endurance hunting may have been another option. Modern humans cannot outrun bovids in a sprint, or even over a middle distance, but in the warmth of the day they can use their ability to shed heat through sweating to run them to exhaustion. So Joseph Henrich has argued that this form of hunting was important in our history, emphasizing the cognitive, technological, informational, and cooperative prerequisites of successful endurance hunting (Henrich 2016).[6] While these debates are intrinsically important, our overall case for the connection between hunting, cooperation, coordination, and communication does not depend on the specific organization of hunting. This in any case is likely to have varied over space and time. All these forms of hunting require cooperation, coordination, and knowledge.

Diet. One theme of hominin evolution since the habilines is an increasing reliance on high-quality foods. As the energetic demands on building and running a human body have gone up (with a slight recent dip), our inbuilt machinery for taking in and processing food has been stripped down, with a reduction in the crushing and processing power of teeth, jaws, and gut;[7] so it is metabolically certain that the quality of hominin food has gone up. Hunted or scavenged meat, marrow, and brain are potential sources of richer and more digestible food. Another strategy is to learn to find and process high-value plant resources, like the underground storage organs (USOs) that many plants in the seasonal subtropics use to get themselves through the dry season. Many of those USOs are difficult to find or difficult to process (as they are mechanically or chemically defended, or both). When it comes to high-value foods, there are no free lunches. Without highly specialized biological equipment, the plants' defenses can only be cracked by some combination of technique, technology, and cooperation (Hill, Barton, et al. 2009). USO harvesting depends on a good deal of social learning, and simple but not trivial technology: digging sticks, processing tools, perhaps fire.

So in arguing that erectines were reasonably regular, reasonably successful hunters, we certainly do not suggest that most of their food was hunted. Amongst ethnographically known foragers living in similar environments, hunting success has very high variance. Most hunts fail (Kelly 2013). Since we conceive of an ambush hunt as involving all or most of the adult and near-adult males in a band, reliance on hunting success would impose episodic starvation on the band, especially given the extra nutrition erectine brains and bodies needed (and very likely, their longer travel times compared to earlier hominins). Indeed, erectine hunting may well have been a luxury good, attempted only when other good resources were reliably available. Richard Wrangham has recently argued, we think very plausibly, that erectine hunting only makes sense as part of a package in which plant foods—bulbs, corms, tubers, and the like—were gathered and processed to make them easy to eat and digest (Wrangham 2017).[8] Unless hunting was restricted to seasons of plenty, hunting probably implies a division of labor. Erectine time budgets could not readily support both a failed hunt and successfully foraging for and processing fallback foods. Again, this suggests a good deal of social learning and a more cooperative life, dependent on coordination, though perhaps in simple forms.

At this point, a skeptical line of thought becomes important. Kristin Hawkes, James O'Connell, and their associates have long argued that the social innovation that made erectine life history possible was female-female cooperation in childcare, and most especially grandmothers giving up direct late-life attempts at reproduction to support their daughters' children (Hawkes 1994; Hawkes, O'Connell, et al. 1998; Hawkes and Bird 2002; Hawkes 2003; Hawkes, O'Connell, et al. 2010). This social and biological innovation enabled forager mothers to wean earlier than chimp mothers (Hrdy 2009), even though the food required by erectine and later infants exceeds that required by chimp infants. The defenders of this view realize that erectines and later hominin males hunted, but they argue that hunts are social signals. Hunters hunt to advertise their capacities. They are not economically motivated, as part of an adaptive division of labor. If this argument is right, we would expect to see hunted resources primarily flowing outside the family group (and without reciprocated return) and significant variation in male reproductive success, correlated with variation in hunting success. Debate continues over these issues.

We agree that reproductive cooperation was part of the revolution in hominin social life, and we agree that erectine life history probably depended on this cooperation; indeed, we argued earlier that some form of reproductive cooperation may well have been essential once hominins were fully, obligately, bipedal. That said, we are somewhat skeptical of this focus on the role of the grandmother and its paired rejection of the idea that hunting is provisioning. For one thing, we cannot assume that erectine mothers had their own mothers and/or other female kin in their residential groups. For it is surely possible that, like chimps and bonobos, erectine groups were characterized by male philopatry, with males staying in their natal groups and females leaving. If that were the case, new mothers would not have lived with their mother, and perhaps with no female kin (see Kaplan, Hill, et al. 2000; Hrdy 2009). We return to this issue in chapter 7. More generally, we argue that reproductive cooperation was part of a more general change in hominin social life, as it transitioned from life structured by a dominance hierarchy, as in the *Pan* lineage and especially the chimps, to the much more egalitarian and cooperative lives of ethnographically known foragers.[9] For grandmother provisioning is a special case of central-place foraging. Grandmothers return to a central place to process their finds (especially if these consist of food for weanlings), thus supporting their grandchildren. Central-place foraging is incompatible with a dominance-structured social environment. For in such an environment, a grandmother is too likely to have her finds taken by anyone who is larger and stronger. The more food she has, and the more she has added value to it by processing, the more tempting the target. Chimps do not return with food to a central place; they feed as they forage. So the picture of the supportive grandmother presupposes at least some shift from a world of aggression, domination, and expropriation to a world of reduced aggression and respect for possession. The reduction in sexual dimorphism in the transition from habilines to erectines may be another sign of this social change, with less male-male competition for dominance. But it is likely that dominance did not just fade away; Boehm and others attribute this change to successful collective action. Indeed, Richard Wrangham has recently made the armed suppression of dominance the centerpiece of his account of human sociality, though he places this much later in hominin evolution than we do (Wrangham 2019). Those at risk from dominance have evolved the cognitive, social, and technical capacities to combine as coalitions, coalitions that constrain those who would be bullies and dominants (Boehm 1999; Bingham 2000; Boehm 2012). Reproductive cooperation was accompanied by greater cooperation in other aspects of hominin social life, including quite demanding forms of collective action; and with greater cooperation comes greater communication.

Life history. These suggestions about cognition, communication, and social learning are reinforced by life history considerations. Erectines seem to be the first hominin species in which we see a distinctive feature of AMH development: childhood (Maslin 2017, p. 32). This is a period in which fairly rapid brain growth and development continue, but in the context of a slowdown in overall physical growth. Erectines had a more rapid life history than more recent hominins; the Nariokotome Boy had finished growing at about 12.3 years, and was just about sexually mature. But there is little doubt that erectines had shifted closer to recent hominin life history patterns, with delayed sexual maturity, relatively long juvenile vulnerability and dependency, perhaps with women often having a significant, active postmenopausal life, and a greater overall lifespan. Slow development to adulthood depends on a relatively long adult lifespan, to repay the costs and reap the rewards of that developmental strategy. Kaplan and colleagues point out that the developmental trajectory of AMH foragers, with their long period of full and partial dependence, could not be sustained if AMH foragers had chimp-length lifespans, which reach a maximum of about 50 years in the wild[10] (Kaplan, Gangestad, et al. 2007, pp. 68–69).

While almost certainly erectine lifespans were significantly shorter than those of AMHs, they were longer than those of most of the large animals with whom they shared the Pleistocene woodlands. As Dale Guthrie points out, such a life history requires low adult extrinsic mortality (Guthrie 2007). In particular, it requires excellent control of the risk of predation. Elephant lifespans are about as long as those of AMHs, and that is possible only because adult elephants are almost immune to predation. Erectine life history did not depend on erectines being quite as predator-proof as elephants, but Guthrie estimates that they would need to be almost as safe.[11] Erectines spent a lot of time in open ground; they were no longer well adapted for rapid tree climbing. Even where tree cover was available, they were no longer adapted to sleep in trees. So how did they stay relatively safe in that environment with its impressive guild of predators?

Perhaps part of the solution was to reduce the risk of encounter by being active in the heat of the day. Most of the more formidable predators were active at night, dusk, and dawn. However, that would increase hominins' dependence on fresh water, and it must be said that other primates do not follow this activity pattern. Fongoli chimps, for example, minimize energy expenditure at the hottest times of day by resting more and traveling less (Pruetz and Bertolani 2009). Perhaps, as Guthrie suggests, at least in their African homeland erectines improved night safety by building barriers of thorn-covered acacia branches around their campsite. This method is still used to protect stock in Africa, as the thorns in question are extremely formidable and are a real threat to the face and eyes of any large animal trying to push through. But it is hard to see passive measures being enough: erectines were large active foragers who had to be out and about a lot. So we infer that these passive measures were supplemented by collective vigilance and information-sharing: information about where predator kills had been made, and hence where predators were resting; these were places for grandmothers with a digging stick to avoid. Perhaps collective vigilance and well-informed avoidance were enough, but we conjecture that avoidance was supplemented by cooperative, aggressive, armed defense. Erectines stayed safe in the bush by moving together, where possible; by staying alert; by knowing about their territory and its dangers and sharing that information; and by mutual support. Even a group of grannies with digging sticks could see off some dangers. One of us (Sterelny) has a traditional aboriginal digging stick. It is a little over a meter long, made of hard, heavy wood, sharp at both ends. If in need, it would be a serviceable weapon. Note, again, that this picture integrates cooperation, social learning, adult-to-adult information-sharing, and flexible use of local resources.

Technology. A new stone tool industry appears in the historical record more or less at the same time as *erectus*.[12] The Acheulian's signature tool is the symmetrical, teardrop-shaped handaxe. Some are extraordinarily beautiful. There is, however, a fair variety of sizes and shapes; not all have the elegance and symmetry of paradigm handaxes. We will go into some detail about the Acheulian stone tool industry and the cognitive capacities needed to make those tools in chapter 5. For now, it is enough to underline their sophistication compared to Oldowan tools. Despite their apparent simplicity, Oldowan tools are not trivially easy to make. But with guidance from a competent knapper, an enthusiastic beginner can learn to knock usable flakes off a cobble in a few hours. In contrast, it takes most contemporary knappers years to learn to make passable handaxes (though admittedly, few spend all their time in this pursuit). The Acheulian industry is much more error-intolerant than the Oldowan, both because each strike has to be more precise in force, location, and direction and (perhaps even more) because success requires a series of well-executed, preplanned strikes. Conceiving and executing the right plan requires a high level of skill. A serious error at any point can derail the whole sequence. Moreover, these planned sequences cannot be stereotyped. There is no simple formula that can be learned and repeated because the correct sequence is sensitive to the initial shape of the core, the raw material from which it is composed, and the qualities of the hammerstone. Further, one's plan of attack often needs to be adjusted in the face of unexpected fracture events. In our view, the lithic crafting of Acheulian tools must be guided by an inner template. The rocksmith must begin with a conception of the shape toward which he (or she) is working, though perhaps, as Steven Kuhn suggests, only a partial conception, with some aspects left free to vary (Kuhn 2020). Moreover, the rocksmith needs a plan to get to that shape (with of course some flexibility, and with details to be decided as the knapping process unfolds).

So erectine lives involved ambush hunting, central-place foraging, perhaps some division of labor and active food-sharing, the preparation of safe sites at which to sleep, some mix of collective vigilance and active defense, and toolmaking. This toolmaking itself required (a) skills that took practice and time to acquire, (b) time and effort to acquire raw materials, and (c) time to work the stone. This implies that their lives were not just lived in the present, with action simply a response to immediate needs. These humans anticipated needs, and invested in the capacities required to meet those needs. The Acheulian toolkit is a further instance of this anticipation, of cognitive horizons beyond the immediate. Moreover, to the extent that effective anticipation of future needs depended on coordination or collective action, erectines were under selective pressure to be able to communicate about those more distant horizons too; that is the focus of section 3.3. In addition, we also think that the Acheulian is further evidence for the increasingly central role of social learning in hominin lives, with perhaps some teaching. Here we rely on an insight of Peter Hiscock (Hiscock 2014). He points out that unsupervised trial-and-error learning of stoneworking techniques is extremely dangerous. The chips that come off often are extraordinarily sharp, and a misplaced blow can send them into the novice's face or hand. Cuts of various kinds are very likely; the loss of an eye by no means impossible. We expect to see teaching (in the broad sense of facilitating social learning) emerge when there is an overlap of evolutionary interests between expert and novice, when the rewards for learning are high but so are the costs, and when their costs can be reduced by teaching at minor expense for the expert (Thornton and Raihani 2008). It is very likely that these conditions were met in erectine groups. Given that, and the intrinsic challenge of Acheulian techniques, we think it is very likely indeed that their transmission depended on assisted social learning of some kind.[13]

These aspects of erectine lifeways intensify as we approach the mid-Pleistocene boundary. For example, by about 1.2 mya we begin to see greater selectivity in raw materials, as well as the emergence of locally distinctive lithic traditions (Shipton 2019). There is also a marked increase in raw material transport distances (Petraglia, Shipton, and Paddayya 2005). Claims about hunting become less controversial, and some of the target animals are now very large (Ben-Dor, Gopher, et al. 2011). Finally, persuasive evidence emerges both for the provisioning of individuals and the provisioning of sites with appropriate raw materials in anticipation of future need (Kuhn 2020).

Distribution. We will briefly supplement our case for a more cooperative, more cognitively and technically sophisticated hominin with one final point. One very striking fact about erectines is their extremely broad and rapid distribution. Within about 200k years of the species' first appearance (from 1.9 mya to 1.7 mya), there are erectines in East Africa, North Africa, Georgia, South China, and Java (Finlayson 2014). The oldest Indian sites are a little more recent, as are the first European sites (in northern Spain). The slightly delayed Indian record is presumably an artifact of preservation. The European dates might genuinely reflect a later arrival. These sites are fairly varied, though none seem to be in closed forest (which is a difficult habitat for hominin foragers) and all are at or near water. But the most important implication of their distribution is that erectines were a very successful species. This distribution is a signal of rapid, perhaps very rapid, population growth that fueled geographic expansion. Whatever their point of origin, and their distinctive adaptive package, founding erectine populations pumped out enough erectines not just to establish themselves in Africa but to spread through much of the tropical, subtropical, and warm temperate regions of the Old World. Of course we cannot prove this, but we very much suspect that this expansion was the result of some combination of improved capacities to plan, improved capacities to manipulate their physical environment, improved social learning, and improved capacities to cooperate and communicate. We now move to the nature of that communication, and to the emergence of wordlike elements.

3.2 Toward Words: Structure

The vervet call system is the textbook example of animal signaling, but vervet calls are not remotely wordlike. While they have some target specificity, they are holistic in two senses. To the extent that the eagle-cough represents indicatively (for it also has an imperative function), it represents a situation or state of affairs. It says *eagle-here-now*. It does not represent any individual, kind, action, or attribute. In that respect vervet calls are more sentence-like than wordlike. Unlike a sentence, though, these calls are also holistic in being unstructured: no element in the cough maps onto a specific element of the situation (or required response). In addition, vervet calls, unlike words or sentences, are stimulus- or situation-bound: both indicative and imperative aspects of the call's function are about the here and now. Moreover, the system is frozen; it is extremely difficult to add new calls.

For a protolanguage to emerge, we need to explain how structured signaling systems evolved in our lineage, in which elements in the signal map onto elements in the situation being described.[14] We need to explain how stimulus- or situation-independent signaling evolved. We need to explain how hominin signaling became readily expandable. We think these features emerged incrementally, for these are not all-or-none traits of a system, and probably to some degree independently of one another.

As we have said, vervet calls and most other animal signals are holistic, and the contrast with signal systems with structure is of great importance. Signal systems with structured signals are not yet systems with syntax, but they are an essential step on the road to such systems. This issue is so important that we devote the next chapter to it, reserving most of our discussion of structure till then. At this point, our aim is to introduce the issue briefly, and connect our views to an earlier treatment by Merlin Donald (Donald 1991; Donald 2001), who also developed a gesture-first account of the origins of language. For us, one important point in favor of a gesture-first view of the evolution of language is that it makes it much easier to see how structured signals evolve. That said, we think the earlier forms of gestural signals were much simpler than those Donald had in mind.

Donald imagined the expansion of hominin communication beginning with something analogous to mime, or to charades, in contemporary life: whole-body gesture, perhaps using props as well. So we might imagine an attempt to convey the location of a specific animal might include some mix of directional gesture (perhaps coupled to some simple convention like repetition or intensity to indicate distance as well) linked to a mime of distinctive body motion, or even a piece of the animal, a distinctive bone or chunk of hide. As vocalization is bought under greater top-down control, mimes can include distinctive sounds. Contemporary foragers are often expert at vocal imitation of the local fauna, and that is an important hunting tool (Lewis 2009), and not just for subsistence foragers. Deer hunters in New Zealand often hunt by calling stags in. They imitate a stag giving a territorial display in the hope that a resident male will arrive to repel the intruder. We think it likely that the importance of vocal imitation in hunting and in mimetic communication helped select for the evolution of elaborate top-down control of our vocal apparatus, control which eventually made the switch from gesture to speech possible. Once it was possible, many factors would favor it. For example, to the extent that communication was commonplace in the context of raw material procurement and the production of stone tools, the hands would have been frequently occupied, making vocal communication more practical (Planer 2017b). Some morphological evidence suggests that top-down control was available to *Homo heidelbergensis*, a species that evolved out of and replaced the erectines; it may have been available to erectines in only a more rudimentary form.[15] We discuss these issues in detail in chapter 6.

At the beginning of this evolutionary transition, gesturing must have been chancy and error-ridden, even though we envisage them beginning (in pre-*erectus* hominins) with communicative attempts much simpler than the mimes Donald had in mind. More ambitious mimes developed from a base of much simpler practices: perhaps indicative pointing, backed by a very simple identifier, when the target of pointing was difficult to spot in bush or woodland. We have in mind something like pointing, then flapping with one's arms to mimic flight to show that the visual target is a bird. Even from such a platform, initially more complex mimes would be difficult to build into a band's social repertoire. Agents and small groups with something to communicate would have to be highly motivated to attempt a more complex message; audiences would need to recognize the mime as an attempt to communicate, and would need high levels of motivation to puzzle it out. As we see it, erectines were motivated to communicate in ways that chimps were not, for they cooperated to harvest resources that were essential for life; chimps do not. If this scenario is approximately correct, there must have been many failures and false starts. Innovations do not stabilize until they work well enough to induce a lifeway change that is self-reinforcing. Initial success might well have depended just as much on associative connections as on an audience correctly identifying the goal of an attempted act of communication. A woman pointing her digging stick at the remnants of a meal of tubers stirs an appropriate memory and inclination. She and her companions go off tuber hunting, but perhaps by her triggering a matching desire, with the interaction not much more sophisticated than a hopeful chimp displaying his erect penis. But once established, enhanced communication would be self-reinforcing. The more it is used, the easier it gets. Once a particular mime has been read and acted on successfully, second and subsequent uses will be easier. Very probably, different mimes will be easier too, for it will be more obvious that they are attempts at communication. Moreover, this will be true not just of the agents directly involved but of the onlookers too. For these innovations in communication are taking place in a social world in which social learning is increasingly important, and where cooperation is increasingly important. So what others do is salient, and worth attending to. Once established and reinforced, we would expect to see conventionalization of the more regularly used gestures and mimes. The general pattern with systems like this is that some iconicity is retained, while time and energy are saved by abbreviating and simplifying displays. For the system responds both to fluent users' demands for ease of use, and to pressures of ease of entry. That is the pattern we see in emerging sign languages in deaf schools and communities, for example (Meir, Sandler, et al. 2010).

A system like this—still organized around iconicity and association—has quite sharp expressive limits. That is true even if the agents using it are adept at exploiting its full range of possibilities. That is not to be expected, especially in the earlier periods of the expansion of communication. The only kinds that erectines could reliably indicate by mime were those with a distinctive appearance, motion, or sound, and perhaps a few that were the target of a distinctive and recognizable kind of activity. In this way a lucky or imaginative Pleistocene hominin could represent, say, a gazelle, by reproducing, by motions of her hand, their distinctive escape bound, or a goat, with a hand sketch of the head motions of a challenging male. Even for the imaginative and the adept, a wholly iconic system will not be able to capture much of the impressive natural history information foragers code in their language. Many species in forager taxonomies have neither utilitarian significance nor a distinctive appearance. Likewise, often there will be no natural way of representing specific places or individuals, except for those with a distinctive gait or physical tic. The expressive power of contemporary languages, both spoken and signed, would be impossible without arbitrariness. But an iconic and partially iconic gesture- and mime-based system, even a fairly simple gestural system, has structure: discrete elements in the mime can represent discrete elements in the environment. The hand mime of the distinctive stott of the gazelle is available to be used in combination with other signs and in other contexts; it and similar items are proto-words: they are available to be used with a constant function in other structured gestures.

3.3 Liberating Communication from the Here-and-Now

All, or almost all, animal communication is about the immediate scene. Human communication is not, and this began, or substantially expanded, with the erectines. The most fundamental driver of this change was the increased decoupling of erectine cognition from the here-and-now.[16] To some considerable extent, they anticipated the future and prepared for it; for example through their caches of stone tools. How did they turn this capacity to think and plan about the elsewhere into a capacity to communicate about it? We suggest that the demands of their new technology gave them some of the cognitive tools they needed. In section 3.1, we pointed out that the emergence of the erectines was roughly contemporaneous with the emergence of a new technology, the Acheulian, that depended on the evolution of technical skills far more elaborate than anything found in great ape lives, or anything known from the material culture of earlier hominins.

The evolution of those technical skills brought long, complex, and precise motor sequences under executive control. Moreover, since those skills are difficult and dangerous to acquire by individual, trial-and-error learning, Acheulian technology selected for improved social learning (as did the natural history skills essential to erectine foraging). That selection made those sequences salient to others, and not just salient: novices were under selection to attend to and parse those sequences, to see the movements not just as a blur but as having structure and organization. This was helped, no doubt, by being able to see the consequences of those actions on stone in its journey from cobble to handaxe. Learning by emulation (attending to the results of actions) complements learning by imitation (attending to the acts themselves). These new technical demands bought complex motor sequences under the control of inner templates rather than external stimuli, and required both improved memory and executive control.

We suggest that these cognitive capacities were redeployed to power communication about matters beyond the immediate scene. For one thing, the acquisition and transmission of Acheulian skills required those hominins to have the ability to take elements of these sequences offline, in demonstration, practice, and perhaps even mental rehearsal. So did the execution of those skills. As we noted in section 3.1, making a handaxe from a cobble requires the artisan to have a mental template of the target, and requires the artisan to use that template to plan a sequence of actions to reach that target. While no doubt for a highly practiced and expert toolmaker much of this is automatized, the variety of starting points (forced by differences in the shape and structure of the cobble), and the need for flexible adjustment as toolworking proceeds, implies that these agents often worked with explicit representations both of the goal and of the structure of the sequence intended to realize that goal (more on this in chapter 5).

This is the cognitive platform for displaced reference using gesture, mime, or demonstration. An agent executing a gestural sequence intending to recruit aid for some plan of action in the future must form an appropriate template to control that sequence of actions and execute it. The same is true of making a handaxe. The agent needs at least a rough conception of the intended material product, and a representation—a plan—of the action sequence that will produce the axe (all going well). As Steven Kuhn points out, in the more developed forms of Acheulian technology, a handaxe cannot be made by successive approximation, with the cobble becoming increasingly similar to a handaxe with each strike (Kuhn 2020). The core must be initially shaped to build striking platforms, before anything handaxish begins to emerge from the stone. Agents must be able to form these action templates without the presence of the usual physical props that trigger recognition. The template must be constructed by free recall from memory, without needing the aid of recognition. This capacity was built or improved by the demands of teaching, learning, and executing difficult lithic skills. Once hominins had the capacity to teach and to learn through practice, they had the ability to initiate an action sequence without most or any of its normal physical triggers being present. When an agent begins crafting a handaxe, stone is certainly present, but its actual shape is very different from the one the stone maker must imagine, to plan her sequence of strikes. Hominins who could execute complex action sequences from memory, without relying on recognition, had a critical component of the cognitive machinery needed to produce a stimulus-independent mime of that activity. For they could produce, say, a sequence of hand actions used to sharpen a blunt handaxe without actually holding a blunt tool or a hammerstone.

To turn a sequence driven by an inner template independent of any actual physical substrate into a communicative gesture, the agent needs a few extra cognitive tools. She needs to combine this template-driven control with appropriate communicative intentions and an improved theory of mind. She needs to be able to represent the goals of other agents, and to represent possible ways her own actions might influence those goals (more on this in section 3.4). At the very least, she needs to be able to predict and represent anticipated responses to her communicative attempts. We have already argued, in 2.4 and 3.1, that the expansion of cooperation in the hominin lineage in the late Pliocene and Pleistocene selected for an improved theory of mind. Ambush hunting, in particular, involves close encounters with large and potentially dangerous animals. In such encounters, each member of the band will need a good understanding of what others intend to do if all goes to plan, and even more, what they are likely to do if something goes amiss. Moreover, they will need some reasonably nuanced understanding of how their own acts will influence what others will do. The same is true if they aggressively defended themselves from predators.

In our view, then, (i) erectine foragers often had cooperative intentions and expectations; (ii) they had template-driven control of action sequences, in particular the capacity to initiate these sequences without the presence of the usual material substrate on which those sequences acted; (iii) they had plans (or information) about the elsewhere and elsewhen that they needed to communicate; (iv) they had theory-of-mind capacities which were an advance on those of the great apes. These capacities were sufficient to intend to communicate, and to recognize communicative attempts. Moreover, (v) the action sequences of others were salient to them. They were apt to focus on them, and to recognize them. As we see it, for agents who were capable of thinking about the elsewhere and elsewhen, stimulus-independent signals are made possible by the combination of inner-template control of action sequences; having and recognizing communicative goals; and an enhanced theory of mind.

3.4 Competence, Comprehension, and Flexibility

One of the most striking and important lessons of the evolution-of-signaling literature is that signals can emerge and stabilize without any intelligent understanding of signaling strategies. Agents can communicate without knowing that they are communicating. Cells communicate with one another, and so do bacteria. That is important, because it shows that successful communication can be one of the resources from which intelligence evolves; communication does not presuppose intelligence. To return to the stock example, vervets communicate successfully about local danger. Do they know that they are communicating? We do not know, but they do not *need* to know. It suffices if the right responses are wired in through some combination of genetic endowment and associative fine-tuning, so that the call is paired to the right danger, and the right response to the right call. In Dennett's terminology, for all we know, the vervets are communicatively competent, but with a minimal (or perhaps no) grasp of the bases of their competence; of why they do what they do. They may have competence without comprehension (Dennett 2017). We suspect that is likely. Competence without comprehension has a price. Automatic, reflex-like communication is cognitively cheap, but also inflexible. Vervets have minimal, or perhaps no, capacity to innovate on the fly, expanding their competence. Such systems can expand, as Scott-Phillips notes, only by lucky accident: an accidental variation in the standard recipe happens to pay off, and that success gets reinforced, through associative learning or natural selection, depending on the source of the variation (Scott-Phillips 2015). In contrast, flexibility on the fly is one of the upsides of competence *with* comprehension. In Sterelny's life as a chess player, the notation for recording and commenting on a chess game was made language-free by replacing language-dependent abbreviations for the name of each piece with a standard icon based on the most common pattern for making that piece, and by adding to an existing list of conventional assessments of moves and positions. In that existing list, "!" meant "good move," "?" meant "weak move," and there were a few others. The Yugoslavian publisher of Chess Informator expanded these conventions, with, for example, "+" becoming "with advantage," "++" becoming "winning," and so forth. Thus a company based in (the former) Yugoslavia was able to sell its chess series all over the world, by a single package of innovations on a system whose basis was well understood by its users. Innovation was guided by comprehension.

Considered from the perspective of vervetese, the linguistic capacities of modern humans are an intriguing mix. In some respects, most of us are competent without much comprehension. We two authors can both recognize a Scottish accent, but neither of us could specify the phonetic differences between Scottish and Australian or New Jersey English. Likewise we, and we surmise most of our readers, can recognize a difference in the following apparently similar sentence patterns. In the second group, the third sentence sounds very strange.

Michael recognized that the move was brilliant.

Michael recognized that **what** was brilliant?

What did Michael recognize that was brilliant?

Michael recognized the fact that the move was brilliant.

Michael recognized the fact that **what** was brilliant?

***What** did Michael recognize the fact that was brilliant?

However, without some explicit training in linguistics, we doubt that many of our readers could diagnose and explain the difference. Many of us have a somewhat vervetish grip on important organizational features of language—how sounds are combined into words; how the position of a sound in a sequence affects its pronunciation; how words are combined into sentences. In contrast, we think that most of our readers have some understanding of the difference between 'Fido' and 'dog'. 'Fido' names a particular dog, whereas 'dog' refers generically to dogs. Here we tiptoe past a voluminous debate in the philosophy of language about the relationship between the meaning of a term and speakers' knowledge of that meaning.[17] For now, we rely on the very minimal claim that, in general, speakers have more of an explicit understanding of the semantic features of their language than they do of its phonology or syntax. We do not think that this understanding is confined to literate, educated societies. Australian aboriginal communities, for example, are acutely aware of the significance of names *as names*, and have important norms about names, especially names of the recent dead.

This difference in comprehension correlates with a difference in flexibility. Most of us would have no chance of introducing into our language an extra phoneme from another language (let alone inventing one). Likewise, morphosyntax. Many languages have "evidentials"; that is, a suffix or affix that attaches to the verb, changing its form, to indicate the supposed evidential status of the report: eyewitness; common knowledge; known via testimony; and so on. In those languages, this is part of the same system that we use in English to mark tense, mood, or number: the so-called "tense-aspect" system. Having evidential status built into a report is a useful feature, but neither of us have the faintest clue about how we might tweak English in this direction. The phonology and morphosyntax of languages change all the time, though typically slowly, and very rarely by deliberate, targeted innovation. Change to the lexicon of a language proceeds very differently. New terms and new names are regularly coined or borrowed, and are added seamlessly and smoothly to the lexical resources of the community. Anyone can do it. Nicknames are invented all the time, though few of them stick.

In one sense, there is no mystery about this difference. If language were designed, this is a feature we would build in, because it is immensely useful to have the flexibility to introduce new terms on the spot, as our environment changes, as we change things in the environment, and as we invent new ways of acting. Occasions on which it would be advantageous to tinker with the organizational features of language arise much less frequently. Moreover, in English (and we conjecture in most or all languages) we have an established linguistic mechanism for expanding the lexicon: "Let's call those things 'snarks'"; "let's call her 'Tilda'." It is also true that "word," "name," and "call" (in the relevant sense) are words of ordinary English; "phoneme," "aspect," "mood," "clause" are more or less specialist terms of art. In virtue of this, the most ordinary English has routine metalinguistic mechanisms for expanding the lexicon. Perhaps this explains why lexical innovation is now so much easier than phonological innovation, but it does not explain how or when the difference arose. We very much doubt that amongst the first words or proto-words were the equivalent of "word" or "name." The metalinguistic machinery that now expresses our partial comprehension of how our language works, and which makes expanding the lexicon so easy, is the result of an earlier and doubtless more rudimentary and less fluent capacity to expand our lexical stock. What cognitive and communicative capacities were required for earlier hominins to grow their protolanguage as needed? We make some suggestions in the next section.

3.5 A Cognitive-Cultural Engine for Expanding the Lexicon

In our view, a group's ability to expand their lexicon at will is the endpoint of a gradual process. As both theory of mind and causal reasoning became more sophisticated in our line, inventing signals would have gradually become more intelligent, more comprehending. To put it another way: it became less dependent on fortuitous coincidence and more dependent on intelligent prediction and interpretation (though we will suggest in section 4.2.2 that the intelligent recognition and exploitation of lucky accidents was also important). Imagine, for example, a sender who wants to draw some individual away from the group to mate in private. Perhaps the sender thinks, "If I wave to her as I walk toward the tree line over there, her attention will be drawn in that direction; and if it is, perhaps she will follow." His plan might fail, of course. But if it succeeds, a new signal may come to exist, something like a "come hither" signal. The receiver may immediately grasp the import of the sender's display, or she may only grasp it in retrospect, or only with repeated exposure. But in any case, assuming she comes to associate the display with the intended request, a new signal will have been established within the dyad. The thing to note is that the creation of this signal would be very unlikely in the absence of the sender's capacity to reason about the receiver's attention, about what she notices, and about the possible, if not probable, consequences of influencing her attention.

Enhanced versions of these capacities would have also increased the rate at which cues turned into signals. When cues become signals, receivers rather than senders lead the way. Imagine, for example, an agent who is attempting to remove some rock so as to make a tool, but failing. This may well be apparent to an onlooker with reasonably sophisticated theory of mind and causal reasoning abilities. Now suppose this onlooker helps out. Then if the actor comes to recognize what has transpired—his striking at the rock caused the receiver to recognize his goal, and to assist him—the sender may well repeat that act when he wants help with this task in the future. In this way, something like a pantomime for stone removal might become established between the pair. Again, without these cognitive abilities on the part of sender and receiver, such a signal would be most unlikely to develop.

We expand on these ideas in the next chapter. For now, suffice it to say that enhanced theory of mind and causal reasoning would intensify rates of signal creation among communicators. But there is no guarantee that newly created signals will escape the confines of the dyads or triads in which they originate. Many of the signals great ape mothers use with their infants, and vice versa, appear to be restricted in this way; they are unique to that mother and that infant (Fröhlich, Wittig, and Pika 2016). Indeed, the use of these signals may even be unidirectional in the sense that B does not make use of the same signals in communicating with A that A uses in communicating with B, even when it is the same message being communicated. It is at this point that the capacity for bird's-eye view representation, discussed in section 2.2, makes all the difference. Recall that this capacity enables agents to represent acts independently of specific agents, and hence enables them to take up different roles in social interaction. Many interpersonal interactions have communicative elements, and armed with bird's-eye view representations, agents can recognize and take up signals they initially experienced as second or third parties. Assuming individuals are in fact motivated to borrow and use successful signals this way, important things follow. First, the signaling repertoire of the group can become somewhat standardized. This simplifies both signal acquisition and real-time production and interpretation. It makes both learning and using signals much more efficient. Second, and more importantly still, it makes signal creation strongly cumulative, as signals created by one generation will tend to be inherited by the next (Planer 2017a). Signal transmission becomes both vertical and horizontal. That is to say, a successful innovation will spread to peers in the same generation, and to the next generation.

The establishment of this basic engine for expanding a group's signaling repertoire would not have required a massive upgrade to the cognitive abilities of great apes; certainly, nothing like the theory-of-mind or causal-reasoning abilities of modern humans would have been required, though bird's-eye view representation is important. A capacity for basic metarepresentation, together with an understanding of simple intentional states (*sees, hears, attends, desires*), was probably enough to kick-start the process. As these abilities were gradually transformed, this engine would have become more powerful still, with a higher success rate, higher bandwidth, higher fidelity. The evidence of erectine lifeways reviewed above makes it very plausible that this engine had been established in erectine populations, and continued to increase in power over the course of the middle and late Pleistocene.

We argued in sections 2.4 and 3.1 that social learning was central to erectine lifeways, and this would have included the more commonly used structured gesture-mimes we described in 3.2 and 3.3. But social learning itself has the same duality as that described above; there can be competence without comprehension. Social learning does not always depend on the novice noticing or understanding what he or she has learned. It can be a reflex-like, a more or less unnoticed result of exposure. There is an illuminating literature on automatic imitation in humans, showing that when people interact, they often begin to match one another in gesture, posture, voice; and the more this happens, the more they like one another. None of this is noticed (Heyes 2011). Probably many norms of social interaction—how close to stand; how loudly to talk—are learned this way. Much or all of early language learning surely is.[18] But obviously some social learning depends on the novice's reflective understanding of the content of what is being learned, and/or the novice's intelligent choice of the expert from whom information is being picked up. Learning a musical instrument from a teacher is often like this, particularly during more advanced stages of learning.

There is major debate within the cultural evolution community about the relative importance of more automatic versus more reflective, deliberative versions of social learning (reviewed in Sterelny 2017). However that debate is resolved, we contend that some form of reflective social learning was in play amongst the erectines. Acheulian techniques require practice—some elements need to be worked into muscle memory. But competence as a whole requires comprehension, for no simple recipe or small set of recipes suffices to turn a cobble into a handaxe. As mentioned earlier, the knapper needs to understand the stone and its possibilities. Some cognitive scientists distinguish "system 1" from "system 2" cognitive processes (Kahneman 2011). System 1 processing is fast, automatic, can be combined with other cognitive activities, is unreflective. Recognizing a familiar tune is an example. System 2 processing is slow, more difficult to combine with other activities, often requires conscious deliberation. Solving a difficult chess problem is an example. This distinction is useful if not treated as a sharp and exhaustive dichotomy. Using it, in our view the acquisition and mastery of developed Acheulian technology depended on some elements of system 2 cognition. So the social learning capacity was there to learn gesture sequences reflectively.

In saying this, we do not imagine highly sophisticated erectine minds. But we think they needed some explicit understanding of their communicative options. Of course, every now and again an agent might try out with success a new signal in hope rather than expectation, as the only option that came to mind. Success then reinforced a lucky break. But to innovate with more than a remote chance of success, in a social setting in which structured gestures were already being used to communicate, an erectine agent—call him Thom—would need to be able to represent possible gestures to himself. That would have been essential for Thom to consider possible signals, and how they might affect the actions of his peers. In terms of the distinction just made between two types of cognitive processing, to introduce new gestures with some facility, Thom needed some system 2 cognitive capacities targeted on his own gestural repertoire. He needed to be able to represent to himself his hopeful "come hither" gesture, perhaps contrasting it with other options: just glancing at his proposed mate (not salient enough); displaying his erect penis (perhaps too salient to others). In addition, to be an adept innovator, Thom had better have had some theory-of-mind capacities, but those capacities need not have been remotely as sophisticated as ours. For Thom's communicative problems, while much more demanding than those of great apes, were still much simpler than those of recent and contemporary hominins. The social world of the erectines was small; their time horizons fairly short; the range of different resources they needed small by comparison to later hominins; their coordination and collective action problems fairly simple. Thom needed to know how to attract attention in ways that got his audience to recognize that he was trying to communicate, or at least that he was trying to influence their behavior. Once he had their attention, he needed some perhaps rough and ready expertise in how to influence their goals toward a small set of familiar actions, events, and things. But that is about all he needed.

3.6 From *erectus* to *heidelbergensis*

The erectines were not just widely distributed. If they were a single species, it was a long-lived one. There were still erectines in Indonesia at 100 kya, perhaps even more recently. But in the heartland of the erectine range, in Africa and Eurasia, at about 800 kya *erectus* was replaced by a still more encephalized hominin, one whose range of brain volumes overlapped with the modern spread. This species is typically regarded as *Homo heidelbergensis*, though as usual there are varying views as to whether the erectines were replaced by a single species, and of the nature of the replacement process. In section 3.1, we attempted to characterize erectine lifeways early in that species' career, and in 3.2–3.4 we explored the cognitive and communicative implications of those lifeways. What happened to those lifeways, and to their protolanguages, as the second half of the Pleistocene rolled on?

We think that the main trend was fine-tuning a lifeway that depended on a good deal of male-male cooperation in ambush hunting; some female-male cooperation in managing the risk of hunting failure (which was surely frequent, especially early in this trajectory); reproductive cooperation, perhaps including fathers; the reduction or suppression of dominance hierarchies; and increased social learning. We strongly suspect that as this lifeway was being established, all of these social patterns were fragile and conflicted: coordination at hunts often miscarried; male-female cooperation in risk management often involved squabbles and conflicts; the suppression of dominance hierarchies sometimes failed; social learning was often inefficient and error-ridden. We envisage a long-term but very noisy trend toward more reliable and less conflict-ridden cooperation over resources, reproduction, and information. Here is why.

One sign is the slow accretion of technology. That is seen in the gradual fine-tuning of Acheulian technology, but also with the crucial new technology of fire. Domesticated fire is surprisingly difficult to identify in the material record, but through Africa from about 1.5 mya there is a scatter of plausible sites, though nothing utterly compelling until about 800 kya (Gowlett and Wrangham 2013; Dunbar and Gowlett 2014; Gowlett 2016; Wrangham 2017). Domesticated fire is enormously consequential, not just for cooking, warmth, and perhaps protection, but also for its social impact, as we shall see in chapter 6. Where fire is shared (and there are great fuel economies in sharing fire for warmth and protection), it trains its users to work, eat, and relax in close proximity while expanding the time available for all those activities. Fire domestication was almost certainly incremental and chancy, often lost, initially involving use and then maintenance of natural fire (common in much of Africa). Even at 800 kya, hominins might not yet have been able to ignite fire at will. Even so, the second half of the Pleistocene saw the beginnings of this second technological revolution. Lithic technologies also show some signs of improved social learning. In an insightful synthesis of the evolution of stone-crafting, Steven Kuhn identifies an important change within the Acheulian tradition, dating from about 800 kya, or perhaps somewhat earlier, with the introduction of prepared core techniques (Kuhn 2020). Using these techniques, the cobble from which the handaxe is to be struck is preshaped, giving greater control over final form. These preshaping techniques become more extensive with Levallois technology, beginning perhaps as early as 500 kya (Wilkins and Chazan 2012; Wilkins, Schoville, et al. 2012).

Another sign is the increased footprint of hunting. By 800 kya, around the origins of *heidelbergensis*, there is no controversy over hominin hunting (Stiner 2002; Stiner 2013). By then, hominins were clearly regularly successful medium- and large-game hunters. There are other signs of improved efficiency, with the appearance of very large game in hominin middens in Africa from about 1.5 mya (Domínguez-Rodrigo and Pickering 2017). Again, around the time of the origin of the heidelbergensians, hominins are beginning to establish themselves in cool temperate areas, that is, areas in which the winter supply of plant food is meager at best. Hunted meat may have been a luxury item at 1.9 mya, so risky that hunting depended on back-up foods that could be gathered easily and quickly. But by 700 kya, in some places it was becoming a reliable source of food, no longer requiring a plan B (Mussi 2007). If hunting had become more efficient, a more reliable source of protein, and if hunters could afford to be more ambitious in their target selection, occasionally taking really large game, then social learning must have become more efficient too. For Binford is surely right to insist that successful hunting, especially with short-range weapons, must have depended on rich, accurate natural history information. Survival by hunting is a very informationally demanding way of life, even for hunters with stand-off weapons. For an impressively detailed set of case studies demonstrating these informational demands, by an author who was both a hunter and an archaeologist, see Frison (2004).

We think fine-tuning these erectine lifeways probably led to two developments in erectine protolanguage. One was an initial shift to a greater role for the vocal channel. By the evolution of the heidelbergensians, as far as one can tell from fossil and genetic evidence,[19] hominin vocal control approximated AMH capacities. Adaptations for vocal control evolve only in an environment in which vocalization is important and control matters: behavior leads, the genes follow (West-Eberhard 2003). So if vocal control was largely in place by 800 kya, we can infer that vocal control was on the increase at some stage between 1.9 mya and 800 kya, probably exploiting whatever developmental plasticity was available. Once top-down control became more available, we think the vocal channel would slowly grow in importance. For as Liz Irvine has pointed out to us, in gesture-based communication, as the number of parties to a conversation increases, the demands on visual attention of tracking others' gesturing rises sharply. Moreover, a gesture-based system significantly constrains the ability to both act and communicate at the same time, partly because of conflicting demands on the use of the hands, and partly because visual attention must be divided between communication and guiding action. Given the importance of communication in the context of tool manufacture and use, this becomes a significant burden (personal communication and Irvine 2016). Finally, and obviously, gesture can be used only when there is good enough light and line of sight. Above we mentioned the transformative role of the control of fire on hominin social life. The control of fire would have extended the social day, and in particular, facilitated communication outside of daytime hours. But hearthside conditions are more conducive to vocal communication than they are to gestural communication; dim, flickering light is not ideal for gesture recognition, but irrelevant to the recognition of sound. We develop this idea in chapter 6. So once voice control is available, efficiency considerations would make it more communicatively important, and we think this transition probably began with the erectines and advanced with the heidelbergensians. Protolanguage was probably always mixed-modality, with some vocal elements. We suspect that over the erectine-to-heidelbergensian transition, those vocal elements became more central.

We also think that the role of arbitrary signs probably grew in this period, in part because vocalized signs are mostly arbitrary. We noted in section 3.3 that as long as protolanguage depended on gesture sequences where each element in the sequence must have some resemblance to its target, the expressive power of the system is quite sharply limited. That would constrain information-sharing about the opportunities and risks in the home range, and constrain expert-assisted social learning. So selection would have favored and led to a less constrained system, if erectines or heidelbergensians had the latent capacities to build and use such a system. We doubt that a shift toward a more arbitrary set of signs would be cognitively problematic. For iconic signs in regular use tend to become abbreviated, conventionalized, and lose their iconic aspects. Admittedly, the experimental data (and data from iconic writing systems) that show these shifts come from AMHs, not erectines or even heidelbergensians (Fay, Ellison, et al. 2015). However, to some small degree, this process of conventionalization and abbreviation is seen in chimp gestures. Regularly used arbitrary signs can be learned by chimps. In an admittedly unnatural environment, Kanzi and Panbanisha learned to use a significant vocabulary of arbitrary signs. So such signs would be less of a challenge for the far more encephalized erectines and heidelbergensians. We return to the transition to the vocal channel, and the increasing role of arbitrariness, in chapter 6.

In chapter 4 we delve into structured signs. We suggest that the earliest form of structure consisted of elaborated pointing, for great apes are almost able to produce elaborated points of the kind we have in mind, and hence hominins only need to be modestly socially more adept to be able to do so. Elaboration is important if points are ambiguous: perhaps because the visual world is cluttered; perhaps because the target is not visually distinctive; perhaps because hominins have come to be interested in discriminating many more aspects of their environment. That third option is important; as we have already noted, erectine lifeways were much more information-hungry that those of the great apes. Ambiguity, we suggest, can often be resolved by adding a gesture (or vocalization). Since these initially depend on resemblance for uptake, point-icon combinations are plausible candidates for the first structured signals.

4 Composite Signs

4.1 Gesture and Structure

One of the most striking features of human language is its complex structure. Phonemes are combined to form morphemes, morphemes to form words, words to form sentences, sentences to form conversations. Animal communication systems, by contrast, tend to be holistic, lacking in structure. There are a few exceptions—most obviously, the songs of some whales and birds and the celebrated bee dance system. But there are profound differences between these cases and human language. In particular, perhaps with the exception of the bees, we do not find such structure being recruited to play a *semantic* role outside of human language. Humans routinely make use of word- and sentence-level structure to encode semantic information. The expressive power of language-like systems in part depends on this structure, allowing agents to combine more atomic elements in novel ways. In this chapter we begin to tackle the emergence of linguistic structure. Needless to say, this is a vast topic, and we cannot realistically hope to address every aspect of it here. In particular, our focus will be on the forms of structure that eventually became the basis for sentence structure, rather than those that became the basis for phonemes and syllables. But in this and the next chapter, we touch on most of those aspects, while singling out a few for more detailed treatment.

We hypothesize that proto-sentence structure was the earliest form of structure. Here we have in mind very simple constructions, ones consisting of two or three signs. We call signs of this sort *composite signs*. Composite signs have other signs as parts, as constituents. These parts are redeployed, or are at least capable of being redeployed, in other communicative contexts. Moreover, they *work together* in some way to specify the meaning of the whole. The parts are not merely two signs that co-occur in space and time. It is this "working together" condition that makes the sign genuinely compositional (Scott-Phillips 2015). It turns out to be surprisingly difficult to give a general characterization of "working together," and so our plan will be to base our analysis on clear examples where the elements are clearly working together.

A gesture-led view of the expansion of early hominin communication leads to a very natural and incremental model of the emergence of structure. At the end of section 1.2, we noted that it was very difficult to develop a credible model of a transition from any rich holistic sign system to a system of structured representations. A gesture-led model avoids this problem by building structure from the bottom up. Except for the very simplest cases, gestures, mimes, and probably iconic representations in general are structured representations. Elaborate forms of pantomime and gesture are inherently structured. If an agent is miming ambushing a horse or picking and eating ripening fruit from a tree, the mime will have a sequential structure (often an activity, the target of an activity, and its results) with elements that in principle could be extracted and reused, for elements in the mime map onto elements in the situation the mime represents. The element of the mime that corresponds to attack from ambush is available to be reused as a mime for ambush hunting with a different target. The simplest and most basic gestures are not structured in this way: an infant chimp raising her arms above her head as a gesture to be picked up; a male chimp reaching out his hand in begging; an indicative point. However, a modest elaboration of these simplest gestures adds structure. If the chimp infant orients herself toward her mother to indicate she is the target of the request that follows; if the agent that points then adds any further element to specify the target of the point (vocal imitation of a bird call, a sinuous motion of his hand to indicate a snake), then the one-element gesture becomes a minimal gesture sequence or mime, with structure. Likewise if a gesture indicating a joint activity is followed by a point, indicating the partner of choice.

What *is* syntax? We take it to be the use of structure at the sentence level to convey semantic information of some kind. To take our example above of the combination of the point and the imitated call, that combination is structured, as the two elements together jointly specify location and identity. But the two signs could be simultaneous or in any order. So there is structure, but not yet syntax, as the physical and structural relations between the two signs carry no semantic information. On the other hand, if vocalizing first in addition meant that the bird had just flown in, that would be simple syntax, as we understand the concept. It is possible to distinguish two types of syntax, linear and hierarchical. As the names suggest, linear syntax involves the use of linear structure—e.g., this word's coming *before* this other word—to convey semantic information. Sentences have hierarchical structure if and to the extent that atomic elements in the sentence are themselves organized into meaningful larger units within the sentence. Thus, in the sentence "The famous Polish professor with one leg was dancing enthusiastically at the party, even though it was on fire," the subsequence "with one leg" is itself a meaningful unit, as suggested by the fact that the "it" in "it was on fire" links back to that sequence. A communication system can make use of both forms of syntax simultaneously. English is like that, in the view of many linguists. Linguists often illustrate the role of structure through examples of ambiguous sentences, where the ambiguity derives from distinct structures organizing the same string of words: one much-used example is the World War II headline "Eighth Army push bottles up Germans," with its two potential hierarchical organizations: [[Eighth Army] [push bottles up Germans]] and [[Eight Army push] [bottles up Germans]].

Syntax will be treated in the next chapter. There we will also address, though in much less detail, the emergence of forms of structure both below and above the sentence level. We will argue that the same cognitive capacities that underwrite human-language syntax also underwrite our use of these other forms of structure. They are all explained by changes in sequential and hierarchical information processing, driven primarily, we argue, by selection for enhanced technological abilities. But more on that later.

4.2 Composite Signs

Animal communication systems rarely[1] feature composite signs, a fact that has often been regarded as a puzzle. One answer has been that the invention of composite signs is very cognitively demanding. More specifically, Scott-Phillips (2015) has argued that in the absence of ostensive-inferential communication, the creation of even a single composite sign is an extremely low-probability event. As explained earlier, on his view, ostensive-inferential communication essentially involves high-level recursive mindreading, tracking of common ground, and more. We think Scott-Phillips overstates the cognitive demands associated with the creation and use of composite signs. It is true that there are some composite signs whose origin is very hard to explain in the absence of considerable cognitive sophistication on the part of sender and receiver. But that is not true across the board. We think great apes probably use very elementary composite signs: an attention-grabber followed by a request, and there is evidence that enculturated great apes can produce somewhat more sophisticated composite signs (Lloyd 2004). The fact that great ape communication is on the cusp of composite sign use suggests that their theory-of-mind, working memory, and causal reasoning capacities suffice, or almost suffice, for the most elementary of those signs.

What might the earliest composite forms have looked like in our line? It is, of course, impossible to know for sure. But we think an especially plausible candidate is a point produced alongside an iconic gesture of some kind. The establishment of such a form would have had profound influences on communication. In particular, it would have greatly facilitated the process of signal creation, expanding the signal repertoire by signal combination. Below we discuss pointing and iconic signs, and then consider how they might have been brought together.

4.2.1 Pointing

Pointing is a human universal. From a very young age, infants frequently point, and do so for a variety of reasons: to share attention, to share emotions, to inform, to request. Great apes, including wild ones, point as well.[2] However, they do so at much lower frequencies in the wild. Moreover, their points are largely limited to making requests, e.g., for food (Tomasello 2007; Bullinger, Zimmerman, et al. 2011); they rarely, perhaps never, point just to share information. Apes can learn to point informatively for one another or for a human experimenter, but this generally involves considerable training (and hence ecologically implausible forms of reinforcement). So we need to explain how pointing became a more regular practice for our ancestors, and how pointing became less tied to its (largely) imperative function. Both questions admit of proximal and ultimate answers. We want to know why, for example, our ancestors found pointing increasingly rewarding. And we also want to know how, if at all, their cognitive capacities had to change as pointing became more frequent and more functionally diverse.

The ultimate question is much easier to answer. As reviewed earlier, there is plenty of evidence to suggest that even our quite ancient ancestors cooperated and collaborated in ways that have no analogue in other apes. They did so in extracting food and other resources; reproductively; and in defense of residence groups, as they moved into a more open, more arid territory full of dangerous predators. In addition, those ancient hominins collaborated in new ways in the context of tool use and manufacture. They were selected to be more tolerant, more prosocial, more willing to help. All of this would have selected for enhanced communication, of both imperative and indicative varieties. As Tomasello (2008) has observed, when two individuals are working together, and one of them has relevant information, it is in the interests of both that that information is shared. The same might be said for efficiently directing one another's behavior. Another possibility, also discussed by Tomasello, concerns signaling one's value as a collaborative partner. As collaboration became increasingly central to hominin lifeways, being chosen as a partner became increasingly important to fitness. One way to demonstrate value is to share helpful information.

The question as to *how* pointing became a more regular, more flexible part of hominins' communication system is more complicated. In contemporary humans, pointing to share information develops very early, and so probably with some genetic basis. Tomasello and colleagues have persuasively argued that at a strikingly young age, human infants develop the desire to share intentional states with others in a way that other apes do not. But we cannot simply project this desire to share information back into the evolutionary past. While the early age at which this sharing-by-pointing develops strongly suggests a genetic basis, we think this basis probably evolved as a consequence rather than the cause of the initial increase in pointing in our line. Selection for such a genetic change (if in fact there was such a change) occurred because collaboration, information-sharing, and cooperation had already taken on a much more central role in hominin lifeways.

We propose that pointing gradually became a more central feature of hominin communicative practice as a result of upgrades to other cognitive capacities. In particular, we think enhanced theory of mind, construed broadly, played a central role. As our ancestors developed a fuller understanding of how others acquire information via the senses; as they developed a fuller understanding of the function that information, or its lack, plays in guiding others' behavior; as they became better at recognizing the information others possessed or lacked, they were more motivated to point, and for a wider variety of purposes. They became much more sensitive to the value of information as a resource, especially as they began to engage in collective action beyond the here and now. If coordinated action takes place in the immediate context, much of the relevant information is perceptually available to everyone present. You do not need to notice or point out what others can see. That changes as common access to relevant information becomes less automatic. Of course, it is not completely automatic even in a shared environment, as not everyone notices everything of importance, as the chimp snake warning calls show. But as the context of interaction becomes more dispersed in space or time, it becomes still less automatic. So sensitivity to information as a resource was enhanced through our ancestors' increased capacity for "mental time travel," as they began to engage more in collaborative activities spread out in space and time. By and large, other apes live in the present. Once our ancestors were capable of thinking about the more distant future, information useful in the future was worth sharing, even if it was not relevant to immediate projects. Hence, one individual might point out to another a location where stone can be sourced for toolmaking, even though toolmaking is not now on the agenda. Felix Warneken, in collaboration with Michael Tomasello and others, has shown that chimps have some prosocial willingness to help when it is easy to do so (see, e.g., Warneken and Tomasello 2006). But they still do not help by indicative pointing. They do not seem to conceptualize information as a resource that others can need. That is why we think a modest enhancement of theory of mind was both necessary and sufficient to get indicative pointing going, given existing prosocial motivations, perhaps even the fairly modest prosocial motivations of chimps and (by inference) those of early hominins.

In sum: pointing took on a more regular and more varied role in our line due to changes in social motivation and social understanding. Those changes were rewarded because of the increasingly collaborative, increasingly cooperative way our ancestors made their living. We suspect this change in willingness to point informationally was driven, at least initially, by enhanced theory of mind, and a greater capacity to detach from the present. In a world in which informational pointing was both widespread and important, selection would then favor genetic changes that made the development of pointing, and of reading points, quicker and more reliable. As elsewhere, in our view the initial changes depended on existing reservoirs of behavioral flexibility, followed by genetic changes that consolidated and extended the new abilities. If we are right, the pointing of modern human infants in part reflects such genetic changes in our line, changes which served to making pointing behavior more automatic.

4.2.2 Icons

Alex wants Ben to know that there is a pig in the bush up ahead, and so she produces an "oinking" noise together with a rooting gesture. Why would she do such a thing? Alex believes that this noise resembles a noise pigs often make, and believes that the motion pattern she makes with her hands resembles a behavior that pigs often perform. Moreover, she believes that Ben will recognize these resemblances. She also believes that Ben will interpret her behavior as communicative, as opposed to some odd display of liberating her inner pig. Ben in fact grasps all of this, and infers there is a pig in the bush up ahead. They successfully communicate.

As we have told the story, both the production and interpretation of this iconic sign involve considerable cognitive sophistication. Alex thinks about how Ben is apt to interpret her behavior, including whether he will recognize her intention to get him to recognize something; Ben, in turn, thinks about Alex's thinking, drawing an inference about what it is she wants him to know. As part of this process, both are led to reflect on what they know together, on their common ground. No doubt our iconic communication is sometimes explained by psychological processes of this sort. Modern, adult humans have at least some comprehension in this domain, not just competence (see section 3.4). But there are many far less cognitively demanding ways in which iconic signs can become established. And these signs can scaffold the creation of other icons by making the *idea* of resembling the intended referent salient to senders. Failure to see this can lead us to greatly overestimate the cognitive prerequisites of creating and using icons. Below we discuss a few of these less cognitively demanding pathways. In many of these cases, the "icons" are more like cues than signals, but no matter: they can still contribute to the beginning of resemblance-mediated communication, as receivers notice not just a specific resemblance, but the value of resemblance in general.

Many of these examples have a common pattern. There are cognitive and cultural processes that are not in themselves forms of communication, but which generate in a focal agent a partial or abbreviated form of an action which both resembles a functional action and can trigger a congruent action in an observer. In many cases, if this is to be the origin of a new signal, reactive intelligence will be called for. One or more of the agents will need to notice that the partial or abbreviated act has elicited a functional response, and thus turn that partial act into a gesture. But anticipatory intelligence of the kind Alex showed is not needed. And there are probably cases where not much reactive intelligence is needed, either.

Imitation. Suppose one individual, Alexi, wants another individual, Barry, to rinse sand off a tuber. Now suppose Alexi knows, on the basis of past experience, that Barry is inclined to imitate Alexi's actions. That would be no great surprise: children are often inclined to imitate adult actions. One thing Alexi might do is simply act out the behavior herself. To the extent that this becomes a recurring situation for the pair, Alexi may no longer need to fully enact the action; Barry may come to recognize a brief dip of the hand in the water as meaning *rinse!* There is no need to appeal to sophisticated mindreading to explain the origins of this sign. Almost all the work is done by Alexi's recognition that Barry tends to imitate her and her expectation that Barry will follow her lead.

Rehearsal. Complex technical skill is associated with the ability to take action sequences offline, in rehearsal. Indeed, it is likely that complex technical skills can only be acquired through practice and rehearsal. The rehearsal of a skill in the absence of its normal physical substrate may become an icon for that activity, or for one of the objects associated with that activity (Sterelny 2016b). For example, we might imagine an individual rehearsing a flaking motion in the absence of a stone core, or rehearsing throwing a spear. In a full-body action like spear-throwing, a smoothly integrated, appropriately timed action pattern is essential for accuracy and power. In both these cases, these rehearsals are aimed at fine-tuning the motor pattern rather than producing any effect on an onlooker. However, through, for example, its resemblance with the act of flint knapping, rehearsal might well arouse a desire to knap in an onlooker (or a desire to hunt, in the other case). That is likely, if the agents in question find successful actions of these kinds intrinsically rewarding. Being reminded of knapping produces positive affect in the knapper. Once either agent becomes tuned to this contagion effect, a rehearsal can be tried out as a signal. In this way, the rehearsed sequence can come to serve as a sign meaning something like *let's knap!*

Impulsiveness. When agents are thinking about certain kinds of actions, motor imagery of those actions often disposes them to perform, in whole or part, the action in question. Cognitive effort is required to prevent this motor imagery from "bleeding through" into their outward behavior (Hostetter and Alibali 2008), but this does not always prevent an outward echo of inner imagery. Hence, we can easily imagine how, in a context where arousal is high, motor imagery might result in outward behavior that resembles the behavior the sender wants to see performed. Think for example of a group of erectines attempting to spook a dangerous predator from a kill, each wanting to drive it off, but waiting for signs that the others will join in. An individual who desperately wants his friend to begin throwing rocks to drive a predator from a kill may well shift in his or her own body posture, becoming side-on with the shoulders cocked. Such "bleeding through" would be more likely in ancient hominins, creatures with less executive control than modern humans. If this does trigger the desired response, again either the agents in question or an observer has the chance of noting this, and recruiting this semivoluntary act as a signal.

Vacuum behavior. Icons might also have their origin in what Konrad Lorenz termed "vacuum behaviour" (Lorenz 1981). This is similar to rehearsal, but here we are dealing with innately driven species-typical behaviors that are normally performed in response to some triggering stimulus (e.g., food). But they are sometimes performed in the absence of the stimulus, typically as a result of the animal having not had occasion to perform the behavior for some time. Lorenz described vacuum behaviors as follows: "In these cases of a motor pattern obviously directed at a certain object but performed in spite of its absence, the naïve observer definitely receives the impression that the animal is hallucinating the missing object" (1981, p. 128). It is not clear to what extent hominins engage in vacuum behavior; we do not have fixed action patterns of the kind Lorenz had in mind. In the human case, vacuum behavior might be a special case of impulsiveness. As desires become more urgent and unsatisfied, motor imagery of the acts that would satisfy the desire become more frequent and more vivid, and "bleeding through" becomes more likely. If the desire is shared, contagion is likely, as with rehearsal, and once again there is the potential to co-opt an initially involuntary and noncommunicative cue as an iconic gesture.

Ritualization. The term "ritualization" refers to the co-option of what was once a functional action pattern as a signal. As the role of the act changes, the act itself is likely to change; it may become abbreviated or exaggerated. This process can be a rich source of iconicity. A paradigm example is the teeth-baring display of a dog or wolf. What is now a signal indicating the threat of an imminent attack was once a preparatory behavior serving to prevent the dog from biting its own lips in an attack. Another clear example of an icon that has evolved this way is the great ape raised-arm gesture, also indicating an imminent attack. This is the great ape version of waving a clenched fist at a foe to threaten violence.

There is a common thread running through these examples: the candidate icon resembles an act either in the sender's or the receiver's behavioral repertoire. That resemblance can trigger, initially without any communicative intent, a congruent response. Once that has happened, there is an opportunity to co-opt the act as a signal. On occasion this might be through simple reinforcement; perhaps more usually, through some more sophisticated route, in which one agent recognizes the causal connection between the triggering action and the congruent response. That may or may not involve the further realization that this causal link between trigger and response depends on resemblance. If that further realization *is* present, it might support generalizing from this example, experimentally trying out further iconic gestures in the hope of an appropriate response. But note that though this experiment would be proactively rather than merely reactively intelligent, it does not require the sophisticated theory of mind we imagined for Alex and Ben.

What about icons that resemble features of other animals or inanimate objects—a sinuous gesture referring to a snake or a pattering gesture referring to rain? Must we invoke considerable cognitive complexity to explain these? These probably depend rather more on proactive rather than reactive intelligence, but even here, while innovation requires some understanding of why existing signals work as signals, that understanding can be quite limited. Suppose that a group makes use of a large number of iconic signs. Then there is a regularity implicit in their communication; many, or perhaps most, of the group's signs resemble their referent in some way. That regularity might be picked up by those learning the sign system, and put to use in subsequent communication. An individual who is forced to create a new sign might take a cue from the other signs he knows. This could result in new icons without the sender having any real appreciation for why signs resembling their referents tend to work. There would be no need for the sender to think about the receiver's thinking, never mind the receiver's thinking about the sender's own thinking, etc. Suppose the group has no standard sign for a hippo, but it does have other iconic gestures, used in conjunction with a point. A sender wishing to indicate the location of a hidden hippo in a swamp might try a point followed by an exaggerated yawn and/or shaping her ears so they resemble hippo ears. To try this, she needs to represent the resemblance of her gestures to a real hippo, and to have picked up on the fact that gestures resemble their targets. But she does not need to know why iconic signs are often understood. Minimally, all she would need to know is that producing a sign that resembles its referent is an effective way of getting one's message across; that is, of influencing the acts of other agents in the ways desired.

At this point, we need to guard against potential misunderstandings. First, the examples above are meant to illustrate possibilities; they are not proposed reconstructions of the actual history of early protolanguage. They are meant to illustrate the communicative potentials that were in play, as hominins evolved theory-of-mind and causal reasoning abilities that were intermediate between those of great apes and contemporary humans. Thom Scott-Phillips has argued that except in the case of rare lucky accidents, composite signs depend on the full richness of human theory of mind (Scott-Phillips 2015), and we suspect many others think the same. For composite signs are clearly very rare in animal communication systems. The examples above show that lucky accidents may well be quite common, and those lucky accidents can be turned into new composite signs by agents who are observant and sensitive to causal connection, but who do not have our theory-of-mind skills. Second, in sketching this set of scenarios, we do not intend to suggest that complex cognitive capacities rarely play an important role in making expandable communication systems possible. A well-chosen iconic sign will satisfy two conditions: first, and most obviously, it will resemble the intended referent in a way that is manifest to the receiver. Second, it will be sufficiently different from other icons that are already in use for other purposes. The second condition becomes harder to satisfy as one's system of icons expands in size. Being able to reason about each other's mental states, and about what is and is not in common ground, will obviously help in choosing signals that satisfy these conditions. We think the development of an expressively powerful and highly efficient iconic communication system probably does require a good deal of cognitive sophistication. But such cognition is not needed to get this process started, and we suspect a much larger body of icons could evolve in the absence of complex theory of mind than is typically realized.

If iconicity helped earlier hominins, why does it play a relatively minor role in the great ape gestural repertoire? In our view, great apes in general and the chimps in particular do not have the right kind of reactive intelligence to build a repertoire of icons in the semi-intelligent way we have been sketching. They are not as good at social learning, and especially not social learning by high-fidelity imitation. Uptake of the iconic signs we have been imagining, in both initial recruitment and regular use, depends on individuals in the residential band paying very close attention to the detailed patterns of arm and hand movements of others. Chimps do not seem to focus at this level of detail; they do not need to. If they do not find salient these fine-grained details of hand movement, they will certainly not notice any resemblance between these patterns and nonsocial aspects of their world, nor will the resemblance trigger an associative connection between a potential icon and its target. In addition, an attenuated capacity for building "bird's-eye view" representations (see section 2.2) is likely to impede uptake. For this form of representation is a bridge between correctly interpreting a signal and using it yourself, when it would be useful. Without it, a receiver may notice a resemblance between a communicative act and target, understand, but fail to reproduce that signal (with similar communicative intent) at a later time, rendering the new icon vulnerable to loss.

4.3 The First Composite Signs

Time to sum up. We have discussed how pointing might have increased in both frequency and flexibility in our line. And we have looked at a number of ways in which iconic signs might become established in a population, given limited cognitive sophistication on the part of senders and receivers. What we now want to know is this: How and why might these signs have been brought together? And what are the cognitive demands of doing so?

There are a number of ways this might have gone. Here our strategy will be to consider one potential pathway in a fair amount of detail. We find this pathway empirically plausible as great ape communicative behavior in the wild is on the very cusp of this form of composite signaling. As such, a small upgrade in the cognitive capacities of our ancestors is all that was required for them to create these composite signs. As the archaeological evidence suggests that our erectine ancestors were distinctly more cognitively sophisticated than great apes, both technically and socially, it is very likely they had the cognitive abilities we presuppose here.

At the heart of this pathway is the process of elaboration in the face of partial communicative failure. Both imperative and indicative communication can feature elaboration. We'll take imperative communication first. Let us imagine a hominin group that now has a body of signs for regulating foraging behavior. An example might be a sign that prompts the receiver to hand the sender a stick or stone which can be used as a tool. Let us also imagine that these signs are iconic, with some sign-target resemblance. So the sign for requesting a stone might resemble the action of hammering; the sign for requesting a stick might resemble the action of digging; and so on. Now let us imagine a situation in which the sender desires a particular stone in the vicinity of the receiver. The receiver recognizes the sender's sign as a request for a hammerstone, but selects a different stone from the one the sender had in mind. What might the sender do?

It is well known that great apes will persist in their signaling behavior when the receiver's response fails to satisfy their desire (see, e.g., Genty and Byrne 2010; Russon and Andrews 2015). What form might our imagined hominins' persistence take? We think this is a place where enhanced theory-of-mind abilities are likely to make a difference. Suppose the sender thinks, correctly in this case, that the receiver has understood the type of object being requested, a stone suitable for hammering. The problem from the point of view of the sender is simply that the receiver has chosen the wrong one. One very natural thing for the sender to do in this case would be to draw the attention of the receiver to the desired stone. The sender might direct attention by pointing at the stone. It is probably not even necessary for the receiver to understand exactly why the sender is now pointing at different stone; all that has to happen is that the receiver's attention is in fact redirected in a certain way. In any case, the upshot (we imagine) is that the receiver exchanges the first stone for the one the sender has pointed at.

Here, the receiver's initial failure to satisfy the sender's desire played a critical role in eliciting the point from the sender. But that might easily change over time. Specifically, the sender might attempt to head off misunderstanding by simply producing a point to the desired stone immediately following his iconic sign. That seems especially likely if this sender is socially "smart," and understands how his point has functioned in similar situations in the past. The result would be one fluent gesture sequence, an icon followed by a point, with no response monitoring in between. Put abstractly, then, the idea is that a modest upgrade in theory-of-mind abilities is likely to have transformed the great ape tendency to elaborate their messages; specifically, we think that elaboration would have become more *intelligent,* taking into account the nature of the receiver's misunderstanding. And it has become more preemptive rather than merely reactive. It is then but a small step, perhaps the *smallest* step, to imagine the sender producing a point to repair the situation, or to guard against the necessity of repair. This whole sequence might then become abbreviated over time.

Were a practice of this sort to become established, the consequences would be profound. First, and most simply, senders might begin to accompany other, already established signs with pointing gestures. Second, the population might repurpose some of these forms, taking advantage of the signs' newly acquired specificity. In a context where another individual is holding a digging stick (and it is understood that they are collectively searching for food), the digging stick gesture, followed by a point to a certain location on the ground might be used to mean *dig here!* Third, new signs might be created that are only intelligible because they are accompanied by a pointing gesture. Here we have in mind signs that are heavily context-dependent. An example would be a general sign requesting that the receiver give to the sender whatever object the sender is now pointing at. In these and other ways, a rich family of imperative signs might come to exist. Table 4.1 provides some more examples, all imagined of course, but whose creation and use are *almost* within the cognitive capacities of great apes.

What about the indicative case? A similar logic applies here as well. Let us imagine a group with a small number of signs for animals they hunt. These signs function to alert party members to the presence of one of these animals. We might then imagine a situation in which one individual spots a monkey, and thus produces the corresponding sign. The sender monitors his peers' reactions, noticing that one of them is still visually scanning the area. What might this sender do? One thing he might do is to simply persist in giving the monkey call, perhaps with increasing intensity.

Table 4.1.
Some examples of an enriched imperative sign system

Form	Meaning
Arm raise + point	*Pick up target of point*
Lower arm quickly + point	*Hit target of point with the object you are holding*
Sweep hands inward toward chest + point	*Give sender the target of the point*
Snapping motion + point	*Snap off the branch that is pointed to*
Cutting motion + point	*Cut carcass at location that is pointed to*

These examples include some rather abstract signs (e.g., the third row). Once more, the idea is merely to illustrate the possibilities available to hominins whose abilities for causal reasoning, theory of mind, and memory were somewhat more sophisticated than those of great apes.

Alternatively—and especially if he has reason to think that it is not a failure on the part of the receiver to understand what kind of animal is being indicating, but rather an inability to *see* it—the sender might point to where the monkey is located. This is not a trivial cognitive feat on the part of the sender. They need to understand quite a lot about the visual experience of others. They might even need to recognize a difference between looking toward something and noticing it. But we need not invoke sophisticated, recursive mindreading to explain it. A rudimentary grasp of the power of pointing to direct attention would be enough.

This process might also work the other way around. That is, a point might be produced for some object, with the receiver failing to detect the object, and giving off obvious cues to this effect. Again, the sender might simply persist in pointing, or he might do something more useful: he might produce a sign that resembles the thing he is pointing at. It is tempting to think that the sender needs a rich understanding of the receiver's mind in order to be motivated to act in this way. We might think, "Doesn't the sender need to understand how his sign will affect how the receiver attends to the current scene?"[3] No doubt being able to reason about the receiver's mind in this way would be very helpful. But perhaps, sometimes, a simpler route might result in signal elaboration. The sender might be given to experimentation, willing to try something, anything. That, together with the fact that he is currently mentally representing the object, might bias him to produce an iconic sign for the object. Alternatively, the sender might draw inspiration from other cases in which icons were used alongside of points. In either of those cases, the sender's additional sign might help the receiver to identify the object without the sender understanding why that is the case.

As above, we might imagine processes of this sort becoming condensed over time, as senders attempt to save time and energy by assisting in interpretation. They might no longer wait for cues that their intended audience is struggling to locate the intended referent before producing an additional sign. As in the domain of imperative communication, the establishment of such a practice would have similar, wide-reaching consequences for the subsequent evolution of the communication system.

Imperative composite signs may well have influenced their use as indicatives, with the former serving as a model for the latter. One reason to think this is that imperative communication probably expanded in our line earlier than indicative communication. Though there are exceptions (Slocombe, Kaller, et al. 2010), very little of great ape communication in the wild is merely indicative, aimed at directing attention to objects in the environment. Hence, the use of composite signs in imperative communication is arguably a smaller step than their use in indicative communication. But in any case, once pointing regularly informed others about environmental states, the range of potential targets would have increased considerably. In turn, that would have created potential ambiguity or uncertainty, and hence in many contexts would have created in turn a genuine need to supplement points with an additional sign.

To repeat: we are not here pretending to describe actual historical trajectories. Rather, our idea is to show that modest increases in theory of mind and causal reasoning, in a context in which agents are motivated to communicate and are rewarded for success, opens up a possibility space for innovation. Much of that innovation, we suspect, took the form of recognizing and exploiting lucky breaks, but we have also tried to show that these opportunities were probably not rare.

4.4 Getting More Complex

We have seen how primitive composite signs—signs bringing together a pointing gesture and an icon—might have emerged. What is a natural next step? Obviously, there is bound to be much uncertainty at this point. But we think there are two natural pathways of expansion: adding an extra sign to extend the sign sequence, and replacing a point with a second icon. Here, too, we think it is probable that imperatives led the way, opening a pathway to somewhat more complex structures. One possible and plausible pathway of expansion would be three-sign sequences that indicate agents, targets, and locations. Recall the discussion in section 3.1 of the challenge earlier hominins faced in finding safety at night, and Guthrie's suggestion that those hominins probably used thorn-tipped acacia branches to build stockades around sleeping places. If that was standard operating procedure, once again we only have to suppose a small increase in sophistication to allow agents to produce and understand a three-part, X-Y-Z sequence in which (for example) the first element indicates the band members being recruited; the second element, the acacia branches; the third, the proposed location of the sleeping site to be fortified. If there are acacias in the immediate vicinity, this might almost be done by three points, but if acacias need to be fetched from nearby, two points plus a mime of (say) sleeping or resting on the ground might become established as a local regularity. One might get similar three-sign sequences with requests to give something to a third party, or to aid a third party in some specific way.

Elaboration is an alternative pathway to a three-sign sequence, with the use of a second icon to draw the receiver's attention to some aspect of the focus of attention, as indicated in a point-icon sequence. Such an elaboration might be needed if there is some particular feature of the referent one wishes to draw attention to—some attribute it possesses or some action it is now engaged in, or some action the sender wishes to recommend. Then the point and one of the icons might work together to designate the subject of the utterance, with the other icon serving to say something about it. Perhaps a snake gesture followed by a gesture of a snake strike to convey the message that the snake is dangerous. So we might get a point, followed by a sinuous snake icon, followed by a strike icon. An elaborated sequence would be useful once senders could no longer expect that merely drawing their audience's attention to some object would be sufficient for them to notice its relevant feature. Here again it is plausible that repairing misunderstanding is a pathway to added complexity. The receiver identifies the subject, but fails to detect the feature the sender is indicating; the sender elaborates with an additional icon; the process becomes condensed. A different kind of complexity would be replacing a point with an additional icon. Put differently, icons might (gradually) come to serve two roles: something like a nominal role and a predicate role. Communication about the elsewhere and elsewhen was likely the main driver for this development. We have in mind here messages such as *the horse is by the river*, produced at a location out of site from the river. With enough common ground, one might communicate this with no more than an icon for horse and an icon for river. A point might be introduced for directionality, if there were two possible rivers, or if the speaker wanted to indicate a more specific location on the river. However, while we sometimes point to refer to absent objects, unless that absence is already very salient to sender and receiver, such points are very likely to be misconstrued. And they would have been even harder to grasp for agents with less sophisticated theory-of-mind capacities than ours.

What beyond this can be said? Nothing with certainty, but there are clues from actual and natural experiments. We can look to language creation in the lab as well as cases of pidgins and emerging sign languages for clues. However, the dynamics of these cases reflect the operation of fully modern minds. Their potential to mislead is obvious if natural selection has equipped us with many language-specific cognitive adaptations, but there is still reason to worry even if that is not the case. We are inclined to view such evidence as on par with ethnographic information in archaeology. Both can serve as a rich source of hypotheses, and can tell us when some independently formulated hypothesis is on the wrong track. But rarely, if ever, can we simply "read off" the past from the present. We thus approach this evidence with caution.

Of particular interest to us, given our commitment to a gesture-led model of language evolution, are cases of homesign. These are cases where a profoundly deaf child lacks any exposure to a conventional sign language, but otherwise experiences a typical social environment. Homesign cases are also distinguished from other cases of emerging sign languages in that the communicative behavior of the child's interlocutors (his or her caregivers) tends to be very simple; it exhibits little or none of the complexity inherent in the child's behavior. We can thus be confident that the caregivers' conventional language is not exercising covert influence on the child's signing.

As homesigning children develop, they employ increasingly sophisticated means for designating different grammatical categories in their signing (Goldin-Meadow 2002). Some of these are quite surprising (to us anyway). For example, Susan Goldin-Meadow reports how David—the child she has studied most extensively—employed lexical, morphological, and syntactic devices to mark out these different categories. She divides David's development into three stages. Initially, David used pointing gestures to designate objects and iconic gestures for verbs and adjectives. So, here, the distinction between nouns on the one hand and verbs and adjectives on the other was marked by the use of altogether different forms. In the next developmental stage, David began using icons for all three roles. However, if he used an icon for a nominal function, he would not use that icon for either a verb or adjective function, and vice versa. (Goldin-Meadow gives the example of David using a laughing gesture to refer to Santa Claus, but never using that same gesture to indicate the act of laughing.) Thus, at this stage, David marked the difference among the categories lexically by only using certain signs for each role.[4] This is especially striking as many of the icons David used as nouns would have worked equally well as verbs or adjectives (e.g., his laughing gesture). Finally, in the third stage, David began using tokens of the same type of icon for multiple roles, but marked the relevant categories using morphological and syntactic devices. For example, when employing an icon as a noun, David would often abbreviate the gesture—he would perform a single twisting motion when using the form to refer to the jar, and perform this same motion multiple times when using it to describe the act of twisting. He also tended to perform an icon in a neutral space (in front of his chest) when using it as a noun and closer to the referent when using an icon as a verb, a form of inflection. At the same time, David would make use of gesture order to mark out nouns and verbs. Adjectives then received a mixed treatment, combining some features of nominal uses (no inflection) and verbal uses (following the noun). So the third developmental stage saw the use of grammar—morphological and syntactic conventions—to mark the distinctions among these different categories. We are not suggesting that David, or any other homesigner, should be taken to be a reliable guide to the stages of language evolution. David might innately possess templates for the categories of noun, verb, and adjective, templates that would not have been present at the inception of language. Moreover, he may innately possess specialized computational procedures for organizing linguistic information that were absent early on. However, David's case and others like it do provide insight into how our ancestors *might* have begun to distinguish grammatical categories, as those categories became increasingly salient to them. Moreover, they illustrate how this might have been done in the gestural domain, without using a spoken conventional language as a model.

The development trajectory of homesign systems such as David's raises a key theoretical question: what drives the emergence of grammatical structure in these systems? What causes them to become more language-like over time? Linguistic constructivists, such as Bates and McWhinney (1982) and Tomasello (2003), see grammatical structure as driven primarily by communicative problem-solving; that is, such structure is produced to facilitate communication between speaker and audience, and is in turn stabilized by communicative success. This hypothesis has been tested using various cases of homesign, and at least in those cases, it would appear to be incorrect (Goldin-Meadow and Mylander 1984; Carrigan and Coppola 2017). If communicative problem-solving were the driving force behind the grammatical development of such systems, then regular communicative partners of homesigners (e.g., family members) should show reasonably high levels of comprehension. We would expect these more elaborated forms of structure to emerge through sender-receiver coadaptation. But that is not what we find. Indeed, considering four adult homesigners, Carrigan and Coppola (2017) found that deaf native signers of American Sign Language (ASL) comprehended the signs of these adults as well as or better (and sometimes significantly better) than the homesigners' own mothers. These native signers of ASL had never been exposed to these homesign systems before, while the mothers had had over 20 years of experience with their homesigning children. In general, it appears that even the primary communicative partners of homesigners tend to understand relatively little of what homesigners attempt to communicate, a fact that homesigners are aware of and to which they are sensitive (Goldin-Meadow 2015).[5]

But if not communicative problem-solving, then what? The main alternative suggestion is that grammatical structure emerges to serve homesigners' own cognitive needs (Goldin-Meadow 2003). More specifically, the thought is that, as the system expands in size, homesigning agents impose order to increase their system's user-friendliness for themselves. The result is structure at both the word and sentence level. This is plausible; once its use has become automatic, it is not difficult to see how imposing a rule on production such as *produce nominal icons in neutral space* would streamline the process. And of course, this view accommodates the finding that grammatical structure emerges even when it does little or nothing to enhance comprehension.

We simply do not know to what extent this process is shaped by cognitive adaptations, language-specific or otherwise, that would not have been present in ancient hominins like erectines. Hence, we should definitely not assume that this process recapitulates evolution. It is possible that David and other homesigners impose order on their system of signs in the way they do only because they possess cognitive mechanisms that have evolved, either in whole or in part, to process linguistic forms. Even if not, it is also possible that their cognitive lives are much richer than those of early hominins, and so their communicative desires, even though they are rarely fulfilled, have driven them to develop systems whose complexity and potential outrun those of earlier hominins. But it is also quite possible that the demands of a sign system on a sender alone—on her memory, on her attention, on her on-the-fly production—select for cognitive strategies that increase its organization, as that system expands in size. If so, as hominin sign systems expanded over time, this became an increasingly important driver of grammatical structure. However, communicative problem-solving situations were surely relevant too. That is especially clear as regards the emergence of shared, stable grammatical conventions. Homesign regularities are not shared and stable. Languages are. No sign system would have been faithfully and reliably reproduced down the generations if it did not earn its keep as a communication system, if it did not enable its users to solve communicative problems. At least in the case of shared grammatical conventions, the constructivist idea must surely be part of the story. However, we do not think there was selection pressure for such conventions until our ancestors were faced with complex communication problems. We discuss the circumstances in which such problems arose in chapter 7. For much of hominin evolution, it was probably unnecessary to specify, linguistically, roles such as agent and patient. And even then, for a long time simple means probably sufficed. We discuss this issue further in chapters 5 and 7.

We have argued that the establishment of a large menu of composite signs was a significant step on the road to evolving language. Great ape capacities for pointing, together with the tendency to elaborate in the face of communicative breakdowns, suggest that signs accompanied by points were an early-emerging and common form of composite signaling in our line. Great apes almost have the capacity for such a sign system. Other theoretical considerations suggest that the signs co-produced with points were likely to be iconic in nature, giving us a reasonably clear picture of the initial expansion of composite signaling in our line, and some plausible models of the further development of composite signs (and plausibility is all we claim). In the next chapter, we turn to the origins of our capacity to make use of these incipient forms of structure.

5 Grammatical Structure

5.1 The Shape of the Problem

In the last chapter we considered the origins of composite signs—signs that have other signs as parts, where the parts work together in some way to specify the meaning of the whole. We suggested that this was probably the first kind of linguistic structure to emerge over evolutionary time. Composite structure was a prerequisite for the emergence of syntax. For us, the simplest form of syntax involves the use of word order, or proto-word order, to encode meaning. This is linear syntax. The English sentence "John saw Jane" means what it does in virtue of the fact that "John," "saw," and "Jane" mean John, saw, and Jane, respectively, but also in virtue of the fact that "John" comes before the verb "saw." Were "Jane" to instead occupy this position, the sentence would mean something different. As we saw in discussing homesign, a version of this can exploit relative position in space, rather than relative position in temporal order, to encode meaning.

In this chapter, we provide an account of the origins of syntax, or, more accurately, of the cognitive capacities on which syntax depends. In our account, the exploitation of linear order evolved earlier than the more complex forms of structure found in living languages. One might be skeptical of this, for not all human languages make use of word order to help encode meaning. The most prominent examples are some Indigenous Australian languages which are described as "non-configurational languages" (Hale 1983). In these languages, roles such as agent and patient are specified entirely with affixes. To see how this works, imagine that "#" is an affix used to denote the agent, and "%" is an affix used to denote the patient. Then the proposition *John saw Jane* might be expressed with "Saw John# Jane%," or with "Jane% saw John#," or in several other ways. So what English and many other languages use order to accomplish, free-word-order languages accomplish with morphology.

There are at least two ways of thinking about such languages. First, one might see them as the original form, with the use of word order being a derived trait in other languages. On this view, the use of order to specify meaning is a late invention. It would not have been a feature of the languages spoken by the humans who left Africa around 80 kya and who went on to colonize Australia. Alternatively, one might think that free-word-order languages reflect the loss of one aspect of syntax in certain language families over time. Here the thought would be that the work that was once done by word order was later taken up by complex morphological structure, making word order epiphenomenal in these languages. We incline toward the latter view. This is in part (though only in part) because we defend a gesture-led view of language evolution. The use of gesture order to specify meaning emerges naturally and spontaneously. That is shown by deaf community and village sign languages, but more significantly it is also shown by cases of homesign. We saw in the last chapter, for example, that the homesigner David spontaneously made use of gesture order to encode meaning. That is not surprising, for gesture often exploits iconicity, and order can be quasi-iconic, with the temporal order of a sequence of gestures mapping onto the causal order of the event being described (or requested). In the ordinary workaday world, sequence matters. It makes a difference whether the snap of a twig occurs before the sound of a bird taking off, or not. So Pleistocene hominins were cognitively primed to notice and respond to the sequential order of events. Likewise, if the speculative but plausible ideas of Shaw-Williams about tracking are right, the spatial structure of the trackway, read by trackers, maps onto the causal and temporal sequence of the trackmaker's movements. This picture is complicated by the switch to the vocal channel, for that channel exploits iconicity much less than gesture and sign. We shall discuss that transition again in chapter 6. For now, suffice it to say that if order had been established as a tool to aid message interpretation with gesture, at least initially, the same tool is likely to have been carried over to voice.

Uncertainties about the emergence of syntax are very great, and our approach in this chapter will reflect an appropriate degree of caution. While in the last chapter we made some quite concrete suggestions about the early format of composite signs, our focus here will not be on the proto-sentences themselves, but on the evolution of the cognitive capacities needed to produce and interpret them. In our view, the empirical ground here is rather firmer, for we think those cognitive capacities were borrowed from ones needed for stone toolmaking. We have a reasonable indication of the techniques needed to make the different kinds of tools found in the record, and of the cognitive demands imposed by those techniques. Those demands have been identified by Dietrich Stout and Thierry Chaminade (Stout and Chaminade 2009; Stout and Chaminade 2012; Stout 2011), whose work we follow here. Stone tools are of special importance, and not just because their traces survive. Ancient hominins undoubtedly made and used wooden tools (Thieme 1997), and if modern chimpanzees are any guide, our ancient ancestors may have used multiple tools in sequential order (Boesch, Head, and Robbins 2009). However, we are very unlikely to learn much about ancient wooden tools. Moreover, wood can do little to shape or otherwise transform wood. Stone, however, can be used to shape stone and other materials, sometimes in very complex ways. Once artisans are equipped with stone tools, other materials can then be worked, though we will usually have only indirect, coarse-grained evidence about those other materials, from residues and wear patterns on the stone tools used for that working.

Our treatment of this topic differs in two main respects from earlier treatments that give a central role to technological evolution. First, we argue that our ancestors already had many of the cognitive capacities needed for simple syntax. Tool manufacture and use explains how an existing set of capacities were upgraded over time, rather than explaining the initial appearance of those capacities. Second, we think our account doubles as an explanation for our ancestors' capacity to make use of other forms of linguistic structure. Long before the last major African exodus, the communicative needs of our ancestors would have forced them to use grammatical devices of some kind to help them communicate. Those needs arose out of the increasingly sophisticated forms of cooperation and coordination that humans were engaged in around the middle-late Pleistocene boundary. Likewise, morphologically and lexically rich languages and protolanguages probably needed to exploit some form of dual coding, and hence of morphemes and words which were themselves composed out of discrete, repeatedly used elements. Without this, it would be difficult to establish standardized recipes for producing and recognizing a large menu of distinct and discriminable lexical items (Dennett 2016, chapter 9). We shall detail this expansion of communicative needs in chapter 7. We shall argue, for example, that late Pleistocene foragers, like historically known foragers, lived in a social world in which residential groups were linked to others by explicitly recognized ties of kinship and reciprocation. These were worlds in which cooperation management worked through the importance of a good reputation, and hence through a reliable flow of information about what others did when they were elsewhere. In short, cooperation depended on gossip, which is communicatively demanding. A gossip needs to be able to specify who did what (including who said what) to whom, where, when, and in what circumstances.

As life becomes more complex because things that matter often happen off-stage, the informational gradient between agents increases, and common ground, while still being very substantial in these intimate communities, shrinks. The universally shared background is no longer as pervasive. So much that could once be left implicit, already known by everyone in the communicative context, now needs to be made explicit. Overtly marking time, agency, and the like becomes important. The grammatical devices that were used might have been either morphological, syntactic, or both.

This chapter has conservative aims: an incremental account of some of the basic cognitive capacities on which the structural features of language depend. In our view, those capacities are not specific to language, nor specialized for it alone. This is not an incremental account of the emergence of syntax itself, charting a progression of increasingly syntactically rich protolanguages until we reach the full equipment of living languages (for an attempt at this, see Progovac 2015, or Hurford 2011). That would require us to commit to a view of what is the best theory of syntax of living languages, as Progovac does. Moreover, we doubt that there was a single pathway. Our best guess is that different hominin communities, under somewhat different selective regimes and from somewhat different starting points, were evolving increasingly complex forms of protolanguage, with varying degrees of independence from one another, and with varying degrees of cross-fertilization (Nick Evans has sketched a scenario of this kind in more detail: see Evans 2017). If that is right, we cannot assume that there was ever a Common Tongue, used by every contemporaneous AMH, nor that all living languages trace to a single common ancestor. We are therefore neutral on whether there is a uniquely correct account of the syntax of language as such, or a unique sequence of increasing organizational complexity in the pathways from ancestral protolanguages to the living languages. Our aim instead is to offer an incremental account of the emergence of the most fundamental cognitive capacities that make possible grammatical structure, broadly understood.

5.2 Baseline Capacities

In seeking to understand the evolution of humans' syntactic capacities, the first step is to characterize the baseline. What capacities did the very earliest hominins possess in this area?

To answer this, we must consider the cognitive capacities of our relatives. We divide this question into two simpler ones, a production question and an interpretation question, and take the latter first.

5.2.1 Interpretation

In chapter 2 we discussed the work of Cheney and Seyfarth on baboon vocalization. While we are skeptical of some of their claims, we agree that their experiments reveal striking sophistication in the way baboons interpret calls. The experiments that are most relevant here concern sensitivity to the order in which calls are heard. The following is a representative example of this set of experiments. Cheney and Seyfarth played sequences of calls of baboons over a loudspeaker. These were recordings of vocalizations of baboons from the same troop as those baboons positioned to hear the loudspeaker. Subjects first heard a "grunt" from one baboon followed by a "fear bark" from another. In baboon society, higher-ranked individuals never fear-bark in response to grunts from lower-ranked individuals, and hence rank relations can be read off from such call sequences. Employing the logic behind the violation-of-expectations paradigm in psychology, Cheney and Seyfarth hypothesized that subjects would look longer at the loudspeaker when the sequence they heard violated rank relations. This is just what they found. Importantly, the effect could not be explained by the novelty of the call sequences in question. For subjects were also played other novel sequences that upheld rank relations, and those sequences did not elicit longer looking-time behavior. Cheney and Seyfarth have carried out a variety of similar experiments involving other call types (e.g., "threat grunts" and "screams") and have found similar results.

Baboons' performance in these cases exemplifies far more than the capacity to distinguish different call sequences; such an ability is doubtless widespread in the animal kingdom. In agreement with Cheney and Seyfarth, we think the best explanation is that hearers interpret what they hear. They have a map of their social world, and they use call sequences to update that map. They represent the identity of the callers and their current rank-order. This information is then available for use in a range of inferences and/or behavioral decisions. Moreover, because hearers react this way in response to completely novel call sequences, we cannot explain their interpretive abilities in terms of past association formation. Their interpretation must be based on piecing together the information carried by each call, together with the order of the calls. This explanation is further supported by the sheer number of call sequences baboons can correctly interpret. A typical baboon troop contains about 80 or so individuals. Thus, as Cheney and Seyfarth point out, the ability to correctly interpret an arbitrarily chosen call sequence implies impressive computational power. We must also add to this the other sequences they interpret, as well as the flexible use they make of all that information. It is very hard to see how this degree of computational power might be realized in the absence of a compositional system of mental representations. With 80 or so individuals and four of five call types, there is a huge number of potential sequences, each of which a typical baboon seems to be able to appropriately interpret. They would require an enormous memory to store each sequence as a holistic unit.

What does this have to do with hominins, though? Baboons are Old World monkeys, and it is believed that Old World monkeys last shared a common ancestor with apes just under 30 mya (Perelman, Johnson, et al. 2011). As we discuss below, apes are also able to make use of the order of signs in interpretation. Hence, parsimony would suggest that this is an ability inherited from the last common ancestor of Old World monkeys and apes. That is not to say that those abilities would have been as highly developed as they are in either baboons or apes. But it does suggest that we have probably inherited from our deep primate past some ability to parse sequences of sounds or acts into parts, and then to interpret the whole as a function of the parts together with their order.

Great ape language learning experiments provide more key evidence. Here we focus on some excellent work by Robert Truswell (2017). Truswell has recently reanalyzed one of Sue Savage-Rumbaugh's corpuses reporting her communication with the male bonobo Kanzi. Truswell's primary aim was to identify the syntactic capacities of other great apes, guided by data from great ape language experiments. These great apes have been exposed to human languages, and hence to the kinds of syntax used in human languages. What they can learn, and what they cannot, tells us about the baseline capacities available as language began to emerge, as well as what was needed to make languages like ours possible. For our purposes, two main points emerge from Truswell's work. The first is that Kanzi clearly makes use of word order in interpretation. This can be seen when Kanzi interprets sentences where the occupant of the agent role also makes sense as the occupant of the patient role, and vice versa. Truswell argues that the correct interpretation of these sentences requires Kanzi to move beyond a so-called "semantic soup" strategy of interpretation. The latter is a deflationary explanation that has been offered for Kanzi's competence in the face of sentences that English speakers would recognize as syntactically structured. The idea is that merely activating representations corresponding to the various words or phrases in the sentence is sufficient to understand its meaning when there is only one plausible way of assembling those representations into a coherent scenario. Consider, for example, the sentence, "Hang the coat on the rack." Since it is not possible to hang the rack on the coat, simply activating representations corresponding to "coat," "rack," and "hang" is likely all that is necessary to act in accordance with the sentence.

Truswell identified a total of 43 sentences in Savage-Rumbaugh's corpus for which a semantic soup strategy would not work. These sentences consisted of 21 pairs where the sentences in a pair differed in the relative order of their parts, and one sentence that was repeated. A typical example of such a pair is given below, including Savage-Rumbaugh's original annotation.

(1) Put the tomato in the oil. (Kanzi does so.)

(2) Put some oil in the tomato (Kanzi picks up the liquid Baby Magic oil and pours it in a bowl with the tomato.)

Truswell then notes that Kanzi acted appropriately in response to 33 of these sentences, for a total accuracy level of 76.7%. This is consistent with Kanzi's overall accuracy across the corpus, namely, 71.5%. Hence, it is very unlikely that Kanzi's behavior in response to these sentences was a fluke.

The other key point in Truswell's work is negative. Kanzi's performance breaks down in cases involving more complex forms of syntax. Truswell illustrates this with sentences from the corpus involving noun phrase coordination (NP coordination). NP coordination is a syntactic device that makes possible arbitrarily complex NPs. For example, in English we often build complex noun phrases using "and," e.g., "the tomato and the oil." The complex noun phrase can then serve as an argument for a verb having only a single argument position. So, (3) is a perfectly acceptable English sentence:

(3) Fetch the tomato and the oil.

Truswell points out that in order to understand this sentence, it is necessary to represent "the tomato" and "the oil" as forming a single constituent. Failure to do so would leave an extra noun phrase in the sentence whose role would be unintelligible (e.g., *Fetch the tomato! the oil*). Thus, if Kanzi does not represent things as speakers of such human languages do, he should exhibit chance behavior in response to sentences like (3). This Truswell characterizes as follows: one-third of the time Kanzi should ignore one noun phrase; one-third of the time he should ignore the other; and one-third of the time he should (serendipitously) respond correctly.

The data more or less conform to this pattern. Twenty-six of the sentences in Savage-Rumbaugh's corpus feature NP coordination. Truswell discards 8 of these trials due to various confounds. For the remaining 18 sentences, Kanzi ignored the first noun phrase on 9 of the trials, the second noun phrase on 5 of them, and behaved correctly in response to the remaining 4 sentences. The sample is small, but this puts Kanzi's accuracy in response to sentences featuring NP coordination at just 22.2%, a stark contrast with his accuracy over the corpus as a whole.

The critical difference between this set of cases and those to which Kanzi responds reliably is that here he must represent the sentences as being hierarchically structured in order to understand them. That is, Kanzi must represent the sentence as having internal structure, with some of the words grouped together to form an organized unit, a phrase, whose meaning must be understood if the sentence as a whole is to be understood. He must represent "the tomato and the oil" as a constituent that has other constituents as parts (see Hurford 2011 for a similar analysis of Kanzi's deficit). Perhaps Kanzi is unable to do this, or perhaps he is simply reluctant because it requires too much cognitive effort (note that even in the simple case his error rate is quite high). Either way, Kanzi's deficit focuses attention on what needs explaining in the case of human syntactic capacities. We need to explain how humans evolved the capacity to fluently make use of hierarchical structure in interpretation.

We shall return to that challenge shortly.

5.2.2 Production

The discussion above suggests that our close primate relatives have much of the cognitive machinery needed to understand simple forms of syntax. What about production? Great apes often produce sequences of gestures for their communicative partners. Such gesture sequences have been closely studied by the so-called St. Andrews team (Genty and Byrne 2010; Byrne, Cartmill, et al. 2017). It is a natural thought that human language syntax might have its origins in great ape gesture sequences, especially if one holds a gesture-led theory of language evolution as we do. However, the current consensus is that these sequences do not contain any form of syntactic structure. Indeed, there is no evidence that the individual gestures work together in a way that would even warrant treating them as composite signs. As a result, gesture sequences have come to be seen as more or less irrelevant to the origins of human language structure. We suggest that this is not quite right.

Pan and gorilla gesture sequences have been studied most closely. Both these apes produce two types of sequences. The first is known as a "gesture bout." Here the gestures occur at intervals of greater than one second, and the sequence is punctuated by response monitoring. These sequences are also characteristically made up of the same type of gesture. The second type of sequence is known as a "rapid-fire sequence." The gestures in these sequences occur at less than one-second intervals. Rapid-fire sequences also typically feature different gestural forms. The St. Andrews team gloss rapid-fire sequences as "streams of synonyms," though it is unclear whether current field methods in fact enable us to tell whether these gestures are really equivalent for the signer.[1] The thought is that rapid-fire sequences reflect a learning process on the part of younger apes. Learners begin by producing a large number of gestures and then work out which gestures are most effective for prompting the desired behavioral response. As they learn this, the number of gestures they produce goes down, eventually resulting in just a few gestures or a single gesture. In addition to explaining the observed age bias, this hypothesis accords with the fact that older apes revert to rapid-fire sequences when placed in a new behavioral context (e.g., when an adult male enters a consortship with a new female). At first blush, rapid-fire sequences would appear to encode a complex message that is a function of the sequence's parts, if not also a function of the order of those parts. With one or two possible exceptions, however, that doesn't seem to be the case. If the research consensus is right, order does not regularly carry meaning in these gesture sequences.

Let us suppose that's right. Still, we think rapid-fire sequences reveal important baseline competences. The rate at which these gestures are produced suggests that at least some of them are represented as a single chunk, as a single unit in the sender's action plan. So in launching a gesture sequence, great apes can chunk a sequence into a single unit, perhaps somewhat as we might represent multiple words as forming a part in a sentence (e.g., a noun phrase). Moreover, senders select specific gestures from among a very large menu of forms, and do so with the aim of satisfying a particular communicative goal. This becomes increasingly clear as the number of gestures starts to decrease. For by this point, the sender is deliberately choosing to perform certain gestures rather than others; he or she is not just selecting gestures at random to try out. These great apes have the ability to produce a specific structured sequence by selection from a range of potentially available sequences. Since receivers are primed to notice sequential order, it would seem that all that is required for these sequences to take on a structured meaning is for the sender's intentions to change.

In sum: great apes seem to have the basic capacities they need for simple syntax. In production, they can recall gestures from memory, choose distinct gestures, and launch sequences of them. As interpreters, if there is information carried by the sequence of calls or gestures, they can learn to recognize and use that sequential information. The step not taken seems to be one of intention, or even of accidental regularization. Great apes do not seem to have slipped into a default pattern that then comes to carry information and which would then be stabilized by success.

5.3 The Easy Problem of Syntax

Above we said that we suspect linear syntax, the semantic use of word order, was likely an early-emerging feature of human language, especially if linguistic communication made extensive and early use of gesture. The main lesson of the last section is this: our ancestors very likely had most of the cognitive equipment needed to use such syntax. So, what had to change?

The first step would have been for composite signs to become a regular part of everyday communication. In the last chapter, we saw how such a transition might have gone, and the cognitive changes that likely prompted it. We also considered some ways in which simple composite signs, a point together with an icon, might have led to the creation of other composite signs, ones featuring multiple gestures, either with or without an accompanying point. Once this happened, the stage would have been set for gesture order to take on a role in encoding semantic information. Point-icon pairs are semantic soups, but once composite signs have become more complex, order can be a useful guide to interpretation. Patterns implicit in composite sign production (e.g., a tendency for the pre-point icons to be used in a nominal role) might be picked up and exploited by language learners. Humans, great apes, and primates in general are good at learning to recognize patterns in data (Saffran, Hauser, et al. 2008). So we hypothesize an initial transition from noncomposite to composite communication, and then from regular composite sign use to simple syntax using linear order.

The use of gesture order or any other grammatical device (e.g., inflection, abbreviation) to encode meaning would have been unnecessary until the communicative needs of our ancestors became more complex. One way to make this idea more concrete is to use the semantic soup strategy from our discussion of Kanzi as an indicator of the degree of complexity at which explicit indicators of role become important. If it is obvious to everyone who is the agent and who the patient, or who is the donor and who the recipient, these features do not need to be explicitly marked. When messages are just about a single agent and their acts or situation, in general the intended interpretation will be obvious. When only one interpretation is plausible, just causing the receiver to think of the right act, agent, or circumstance is probably all that is necessary to get one's message across. But as messages become more complex, if and as they come to involve multiple agents and/or sequences of acts, just specifying the acts and the agents will leave too many possibilities open. Hence, we suspect that it was only once a semantic soup strategy became unreliable that syntactic and/or other grammatical devices emerged. Importantly, merely getting others to attend to the right agents would rarely work well in messaging about the social domain, once a certain degree of role flexibility has appeared. Likewise, in general, when communicating about the actions of two (or more) individuals, it will make sense for either of them to have been the agent or the patient. For individual behavioral repertoires will typically overlap substantially. Moreover, given the foundational role of such communication in shaping cooperation and collaboration, the costs of miscommunication in this domain can be steep. It is very important to know whether it was Bosco who failed to repay Frodo or the other way around.

This is methodologically encouraging. For it means we can connect the evolution of grammatical devices like linear syntax to archaeological traces of enhanced forms of cooperation and collaboration (see chapter 7). It is of course possible that grammar emerged even earlier and/or in response to other communicative needs of our ancestors. But we do think that communication about social events, actual and possible, must have been a key driver. Cooperation typically requires coordination, and the more spatially and temporally extended the cooperation, or the divisions of the profits of cooperation, the greater the communicative requirements. Once again, we will leave detailed discussion of this to chapter 7, but as a sample of what we have in mind, consider the late Acheulian site of the Qesem Cave (see Stiner, Gopher, et al. 2011; Barkai, Rosell, et al. 2017). From the dating, these were probably heidelbergensians. The basic procurement strategy at this site was probably some form of ambush hunting (with a bias toward prime adults) combined with gathering. This is a form of collective action, but the profits of collective action were not divided and consumed on the spot. Rather (judging from the surviving bones at the site), the kills were disarticulated, and parts not worth the transport costs (e.g., hooves) were discarded. Other parts were transported back to home base. The flesh and bones were cooked, probably by roasting, at a very large and hence communal hearth. The bones were broken open and the marrow consumed; likewise the meat was divided and eaten (judging from the scatter fragments of the gnawed bones). This cave site records delayed yet communal sharing. Kills were divided for transport back to home base rather than divided for immediate and individual consumption. This is an archaeological snapshot of orderly and sustained cooperation. It is clear that the communicative demands of this kind of communal life, whatever they were, considerably exceed, say, those generated by confrontational scavenging as a party of hominins collectively drive a carnivore from a kill, and then individually scrounge chunks of whatever remains. Rough-and-ready cooperation by such armed mobs was very likely a critical evolutionary precursor for the more orderly teamwork that the Qesem Cave seems to record. But just as the cooperation was rough and ready, so, very likely, were the requirements of communicating to coordinate.

In our view, neither the evolution of our capacity to use linear syntax, nor the actual creation of linear syntax in our line, pose deep mysteries. The great apes have the capacity, or almost have it. The main changes were changes in the social and economic environment that selected for the exercise and enhancement of these latent capacities. Where there seems to be a large gap between us and other animals, including our close primate relatives, is in the capacity to use hierarchical structure in communication. But what, more precisely, does hierarchical structure amount to? This, it turns out, is not a simple question.

5.4 Hierarchical Structure: The View from Linguistics

The dominant conception of hierarchical structure is conceptually linked to formal grammars. This is the notion in play in linguistics, as well as in most evolutionary discussions that draw on linguistic theory. The sentences of a language have hierarchical structure if that language is appropriately specified by a certain kind of grammar, specifically a context-free grammar. Such grammars are seen as more powerful than regular grammars, the simplest kind of formal grammar. (See boxes 5.1 and 5.2 for the technical details of regular and context-free grammars.)

Intuitively, the idea is that only context-free grammars can build sentences, or "generate strings," as linguists would say, that have parts in the way we ordinarily think about human language sentences as having parts. English speakers, for example, recognize "John" as a part of sentence (4):

(4) John walked.

This kind of structure cannot be captured with a regular grammar. A regular grammar specifying (4) would necessarily have two constituents, "walked" and "John walked," as every constituent must, in virtue of the form of the production rules of a regular grammar, extend to the end of the string. As a result, regular grammars cannot generate strings that have *hierarchical* structure, which is a special type of internal structure. Regular grammars cannot produce strings that have constituents inside other constituents (except in the degenerate sense shown in figure 5.1). Consider sentence (5), for example. English speakers recognize "with the dog" as part of (5), but also "the man with the dog" as a part of (5).

(5) The man with the dog went home.

Box 5.1

Formal grammars

A formal grammar is a finite set of production rules that generates a (formal) language, a set of strings. Strings are formed on some alphabet of terminal symbols. Terminal symbols contrast with nonterminal symbols, which can be rewritten as one or more other symbols; terminal symbols cannot be rewritten. When formal grammars are proposed as grammars of natural languages, terminal symbols are words or morphemes. Production rules relate symbols to symbols; they license us to rewrite the symbols on the left side of a rule with those on the right side of a rule.

Regular grammars

Formal grammars are divided into classes depending on the form of the production rules they contain. The simplest class of grammars consists of the so-called "regular grammars." In a regular grammar, every rule takes the following form: the left side of a rule has a single nonterminal symbol, and the right side has either (a) a single nonterminal symbol and a single terminal symbol, (b) just a terminal symbol, or (c) the empty string (ε) (which for all intents and purposes can be thought of as another terminal symbol). Examples of each type of rule are below:

(5.1.1) a. X → aX

b. X → a

c. X → ε

Regular grammars generate strings that are said to have linear structure only. Every constituent of a string generated by a regular grammar necessarily extends to the edge of the string, where a constituent just is any substring that corresponds to an expansion of a nonterminal symbol using the production rules. Which edge this is depends on whether the grammar is a right regular or left regular one. In the case of a right regular grammar, a nonterminal symbol on the right of a production rule is written to the right of the terminal symbol, as in rule (5.1.1a). This results in constituents that extend to the right edge of the string. As an illustration, consider the string "aaa" generated by applying rule (5.1.1a) twice followed by rule (5.1.1b). This string contains exactly three constituents, namely, "a," "aa," and "aaa," moving from the right edge of the string to the left (figure 5.1). In the case of a left regular grammar, a nonterminal symbol on the right side of a production rule is written to the left of the terminal symbol. This results in constituents that extend to the left edge of the string.

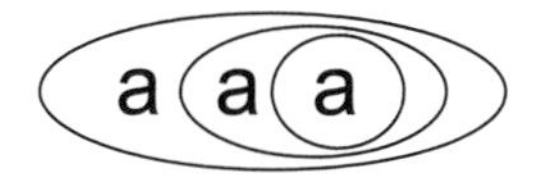

Figure 5.1

This edge constraint can also be illustrated in tree format (figure 5.2). The linear structure of the string shown in figure 5.2 is reflected by the right linear growth of the tree. No string generated by a (right) regular grammar can have a branch that is longer than the rightmost branch. For that would imply a constituent whose boundaries do not extend to the right end of the string.

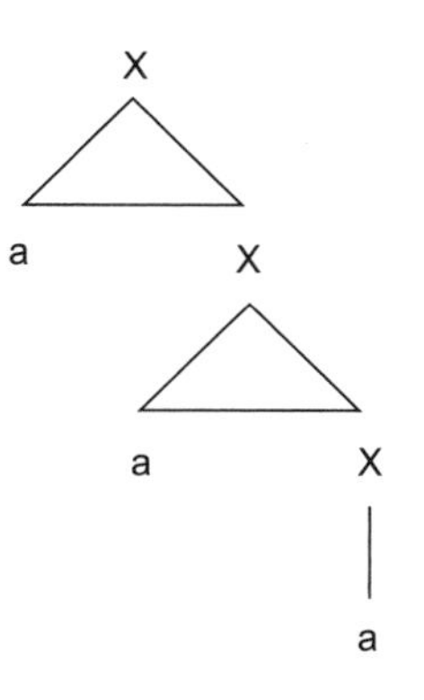

Figure 5.2

In more intuitive terms, this constraint implies that regular grammars cannot generate strings that have internal structure, or have internal structure only in a degenerate sense (i.e., the kind of structure shown in figure 5.1).

Were we to draw a tree diagram corresponding to the intuitive structure of (5), it would look something like figure 5.3. Notice that there is an intuitive sense in which this constituent can be further expanded, as in:

(6) The man with the dog with the black rear leg went home.

In sentence (6), "with a black rear leg" seems to belong with and be part of the whole constituent that begins with "the dog." In contrast, and again intuitively, in:

(7) The man with the dog and the bottle of whiskey went home

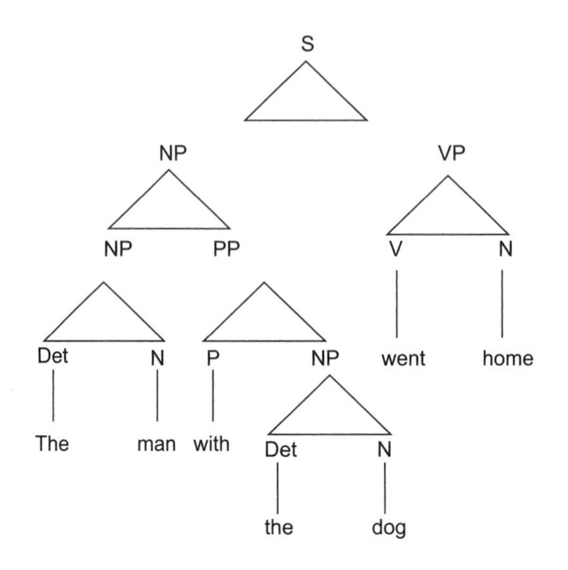

Figure 5.3

"and the bottle of whiskey" is not part of the constituent that begins with "the dog."

The intuition that "with the dog" is a part is reflected, from a formal grammar perspective, by the fact that each of these words is dominated by a single nonterminal symbol, namely, a PP (prepositional phrase). And the intuition that "the man with the dog," is also a part is reflected by the fact that each of these words is in turn dominated by a single nonterminal symbol, namely, an NP (noun phrase). Note that this tree has several branches that are longer than the rightmost branch.

To capture these kinds of structure, it is necessary to move to a context-free grammar. These grammars can contain any production rule a regular grammar can contain. But they can also contain rules that enable us to build (or recognize) complex constituents inside the sentence. Regular grammars can only build greater complexity—"expand the string," in the jargon—at one end of the string.

We agree with the linguists that a sentence has hierarchical structure when it has parts that in turn have parts: when it has a nested part-whole structure. We agree that this is intuitive; "with a dog" seems to be a part of sentence (5) in some important sense, and in that same sense "dog went" is not a part of (5). But this buries a very important question. For in an obvious sense, "dog went" *is* a part of (5). It is a physical part. So we need some

Box 5.2

Context-free grammars

Unlike a regular grammar, a context-free grammar can contain production rules that feature multiple nonterminal and/or terminal symbols on the right side of a rule. (The left side, however, is still limited to just a single nonterminal.) This difference makes all the difference. It is what allows for the building of strings that have constituents inside the sentence, where these constituents can, in turn, contain other constituents. Below is an example of a context-free grammar that would generate a string with the structure shown in figure 5.3.

(5.2.1) a. S → NP VP
b. NP → NP PP / Det N
c. PP → P NP
d. P → with
e. VP → V N
f. Det → the
g. V → went
h. N → man / dog / home

Note that "/" is used to indicate a separate rule. So rule (5.2.1b) is actually two rules, one rewriting NP as NP PP; another rewriting NP as Det N.

For any string that is generated by a context-free grammar, there exists a regular grammar that generates that string. This follows from the simple fact that we can always write down a series of production rules that will generate the linear order of a string. But equally important is the set of strings that a formal grammar *keeps out*, that it precludes. It is trivial to specify a regular grammar that generates the string "The man with the dog went home," and other strings that English speakers would recognize as sentences of English. The trick is to specify a grammar that generates all and *only* these strings. A grammar of English should not generate strings such as "The man with the dog went home the man." This cannot be done without production rules that permit us to rewrite a nonterminal symbol on the left side of a rule with two nonterminal symbols on the right. Hence, English is (minimally) regarded as a context-free language.

Recursion

Recursion is often confusingly described as being made possible by a context-free grammar. Technically, recursion is a property of a derivation (i.e., a sequential application of a formal grammar's production rules, resulting in some symbol string). A derivation is recursive when it makes use of the same production rule more than once. We can then say a grammar is recursive when it

Box 5.2 (continued)

allows derivations of this kind. It is via recursion that an infinite set of strings can be generated from finite means.

Regular grammars can be recursive. In fact, the grammar shown in rule (5.1.1) is recursive, as the rule X → aX can be applied over and over in the context of a single derivation. However, some theorists reserve the term "recursion" for a stronger property, namely, the ability to expand a string from the *inside* by embedding constituents inside of other constituents. This is known as "center-embedding." It is true that a context-free grammar is needed to specify constituents that are center-embedded. For this requires a production rule whose right side contains a nonterminal symbol flanked by two other symbols (e.g., X → aXY or X → aXa). Hence, on this notion of "recursion," there is indeed a set of connections holding among context-free grammars, recursion, and hierarchical structure. Theorists who use "recursion" in this way typically describe the kind of recursion regular grammars can support as mere "iteration" (see, e.g., Corballis 2014).

account of what counts as a part. That account should explain why "with a dog" is a part, and "dog went" is not; it cannot just recycle the intuition. The answer from linguistics appeals to the formal grammar that generates—builds—this sentence and of course the other ones in the language: "with the dog" is a part of (5) because (5), the thought goes, is generated by a formal grammar that treats all of these words and only these words as a value of a single nonterminal symbol (a PP). In a representation of the structure, those three words and only those words map onto the PP symbol. And "man with the" is *not* a part of (5), and nor is "dog went," despite obviously being physical parts of the sentence, because they are not ways of filling in a single nonterminal symbol in the grammar generating this sentence.

We do not think this answer can possibly be right. Firstly, it gets the order of explanation the wrong way around. A certain class of formal grammars provides the right grammar, a member of this class, to describe the language we speak because of facts about how we use and understand our language. Prima facie, we need a cognitive explanation of what a constituent is, one built around language production and interpretation. Such an explanation should both explain our intuition that "with a dog" is a part of sentence (5) and "dog went" is not, and also explain why a context-free grammar is the right way of specifying our language. In this regard, it is worrying that in contemporary linguistics the psychological standing of formal grammars is at best unclear. It was once common to think that a formal grammar described a psychologically real form of structure, one that explained special features of our language capacity. Humans are capable of producing and understanding an infinity of sentences. Moreover, the sentences we judge as grammatical obey quite specific constraints. We do all this despite having finite computational resources. How? By "having knowledge of" a certain formal grammar, the thought went. Now very few theorists think that we can read off the psychological processes that explain our capacity to use our language from the kinds of formal grammars linguists produce for natural languages. For one thing, these grammars, despite the talk of building or generating sentences, are supposed to be completely neutral between language production and language comprehension. But it is hard to see how the process of understanding a sentence could be the same as the process of deciding on and producing that sentence. It is one thing to build a structure, another to recognize the organization of that structure. For that reason alone, the actual steps taken by an English speaker in producing or understanding sentence (5), for example, may bear little resemblance to a derivation of (5) using the kind of rules shown in (5.2.1a–h).

If all that is required of a grammar is that it provide an elegant and compact description of a given language for the linguists who study it, then we can indeed regard grammars as saying nothing about the mechanisms of production or comprehension. Perhaps that is the view of some linguists. However, on that approach, it is unclear what becomes of a grammar-based account of hierarchical structure. Once we give up the idea that formal grammars specify (perhaps at some level of abstraction) the computational processes involved in building or understanding utterances, it begins to seem as if the claim that some string has some particular hierarchical structure is a matter of descriptive choice. The string has such structure simply because that is how we as theorists choose to analyze and represent it. It would be somewhat akin to the way the rings in a cross-section of a tree carry information about its age and growth conditions. Those ring structures carry information for dendrochronologists, using trees to date environments of the past. But they do not carry information for the tree. As far as we know, trees cannot consult this archive of their former growth patterns, notice that a dry year is typically followed by a few more, and so plan on the coming growing season being dry. Analogously, Stefan Frank and colleagues in a recent skeptical review article about hierarchical structure in language argue that treating sentences as hierarchically organized is useful for linguists, enabling them to give compact accounts of the kinds of sentences found in particular languages, but that those structures are not used in speaking and interpreting (Frank, Bod, and Christiansen 2012). We are inclined to agree if hierarchical structure is defined in terms of formal grammars, and if there is no commitment to the cognitive reality of these grammars; that is, if they are not accounts of how languages are used. More on this at the end of section 5.7.

Secondly, hierarchical structure in language is not limited to the sentence level. Most obviously, it is also visible below the sentence level. Words are composed out of morphemes, which are themselves composed out of phonemes. This is an instance of recursive part-whole structure too. It is also possible to see hierarchical structure above the sentence level, and some theorists think this is an important kind of structure (e.g., Levinson 2013). A conversation can be understood as composed out of shorter bits of dialogue, which are themselves composed out of sentences. So there is a sense in which language is hierarchically organized from top to bottom. Defining the hierarchical structure of a sentence by appeal to the formal properties of the grammar that describes the language's sentences sits awkwardly with this view of language. For it gives up any prospect of coming up with a general answer to the question "What is a part?" We think the view from neuroscience might help.

5.5 Hierarchical Structure: The View from Neuroscience

Neuroscientists standardly conceive of actions as exhibiting part-whole structure. The act of making breakfast, for example, is seen not as a single act, but rather as a complex sequence of subacts. The preeminent nineteenth-century psychologist Karl Lashley is generally credited with this idea. Lashley pointed to errors of omission as evidence for such structure; in making breakfast, one might forget to take the milk out of the fridge, for example. If our *make breakfast* routine was stored as a single ballistic sequence, once launched we would see the whole action stream unfold, with variation occurring as just noise; we would not see elements omitted, added, or reordered. Lashley took the fact that extended action sequences are not ballistic to show that our action representations must be internally structured and composed from simpler elements, making it possible to skip over, add, or change the order of one or more components of an act.

As in the linguistic domain, hierarchical structure in action is understood as nested part-whole structure. But here we again encounter the question of what counts as a part. Neuroscientists answer this question in several ways. One approach takes its lead from linguistics. It sees action as generated by a formal grammar—an "action grammar" (e.g., Jackendoff 2007; Stout 2011; Boeckx and Fujita 2014). The structure of an act string is imposed by the formal grammar generating the string. We are suspicious of this approach, as it runs into the same issue concerning psychological reality. Unless it is literally true that agents derive their actions from an action grammar, then it is far from clear what the notion of hierarchical structure amounts to. In particular, it is not enough for actions to be usefully and systematically described in this way. Since we can both act and recognize actions, as with language, this "action grammar" would have to be neutral between production and recognition. While we do not say this is impossible, we would like to avoid this bet if possible.

There are two approaches independent of formal grammars. One identifies the parts of an act string on the basis of instrumental, causal relations holding among the physical parts of the string, of the actions themselves (as distinct from whatever neural systems control those actions) (e.g., Botvinick 2008). So, for example, *open the fridge* and *take out the milk* are parts, proper constituents, of a larger action string containing them because opening the fridge makes it possible to take out the milk. Instrumental relations can hold among parts that are themselves complex, and when that happens, we have a case of hierarchical structure. The action string *make breakfast* has many parts and itself makes possible the action string *eat breakfast*. These parts, in turn, are further parts of the action string *get ready for work*. Some neuroscientists prefer this behavioral approach exactly because it makes no mention of how the agent mentally represents an action string. Whatever its merits in other contexts, this behavioral-causal notion of a part does not apply well to language, for it probably counts all the physical parts of the sentence string as constituents; saying "dog went" is a causal precondition of saying "home" in our above example. So it would give the wrong specification of the structure of "the man with the dog went home." In contrast, an alternative approach identifies hierarchy, and hence part-whole relations, by appealing to the way agents represent action—their own and others'. In particular, this account turns on how agents represent the relations between goals and subgoals (e.g., Badre and D'Esposito 2009; Badre, Hoffman, et al. 2009; Christoff, Keramatian, et al. 2009). Roughly, it says that a substring is a part just in case the agent has a goal representation corresponding to that substring. This conception of the structure of action is associated with a far richer body of neuroscience research. For that reason we prefer it, as it keeps its eye squarely on the underlying psychology. Importantly for us, in contrast to the behaviorally defined notion, this sense of hierarchy extends naturally to language. For producing a sentence is executing a complex, and in our view structured, sequence of actions; interpreting a sentence requires receivers to identify that structured sequence.

We think these ideas about the representation of goals, and how those representations are interrelated, are very promising, though they are expressed in a rather opaque terminology. The literature in question describes goal representation as more or less "abstract," in two senses of abstractness. Representations can be inactive or active: causally salient in directing action or interpretation, or not. *Temporal abstraction* refers to the duration of activation, with more temporally abstract goal representations being active longer. In the course of getting ready for work, the representation MAKE BREAKFAST is active for longer than the goal OPEN FRIDGE. The other type of abstractness is *policy abstraction*. This refers to the number of goal representations a given representation controls. MAKE BREAKFAST is more abstract than OPEN THE FRIDGE in this policy sense too. High degrees of temporal abstraction typically coincide with high degrees of policy abstraction, and vice versa. However, it is possible for the two to come apart. A goal representation might control a large number of lower-level goals, but be active for a relatively short period of time if the latter are all swiftly executed. Conversely, though it would seem less commonly, a goal might be active for a long time while controlling relatively few lower-level goals, perhaps because a subgoal simply takes a long time to execute. GRIND THE EDGE UNTIL IT IS SMOOTH AND SHARP might be such a subgoal. For our purposes, policy abstraction is most relevant, for this is where the connection with hierarchical structure lies. Talk of "abstractness" in what follows should be interpreted as referring to policy abstractness unless otherwise noted.

We read neuroscientists working in this paradigm as hypothesizing something like the following computational system. Here we work with a relatively detailed characterization, though this is primarily to aid in clear exposition; the exact computational specs of the system are actually not crucial to our argument.

(i) The system contains a multilayered spatial array that it uses as a workspace. Representations on the same level have a sequential order. A representation at a higher level controls one or more representations at lower levels.

(ii) In assembling an action plan, a goal G is selected. This constrains the selection of other goals. Specifically, goals whose sequential completion would satisfy G are selected. This process is repeated until basic-level goals have been selected. These representations are written to the array so that higher-level goals control the lower-level goals that are aimed at satisfying them.

(iii) A given representation gains causal efficacy over the agent's behavior at a time by being activated. Activation spreads both vertically and horizontally through the array. Activating G causes the first goal that G controls to become active. This process continues until a basic goal is reached. Once that happens, the basic goal is executed and deactivated. Activation then spreads horizontally to the next goal that is controlled by the same higher-level goal at the preceding level. Once all of these goals have been executed, the higher-level goal is deactivated and activation spreads horizontally to the next goal that is controlled by the same higher-level goal at the preceding level. And so on. (See figure 5.4.)

(iv) Sometimes a lower-level goal cannot be executed or execution fails. When that happens, the higher-level goal controlling it at the preceding level remains active, while the system searches for another goal (or sequence of goals) to replace the frustrated goal. If no such goal (or goals) can be identified, a replacement is sought for the higher-level goal, and so on. The logic here is that the system will change the fewest possible components to create a functional action plan. It does not start over from scratch upon encountering a problem.

This general picture has neuroscientific credibility. fMRI and other neuroimaging techniques have localized these operations to prefrontal cortical areas. In particular, Broca's area (area 45) plays a critical role. This region has been observed to systematically spike upon the completion of both higher- and lower-level goals, suggesting that it is involved in switching

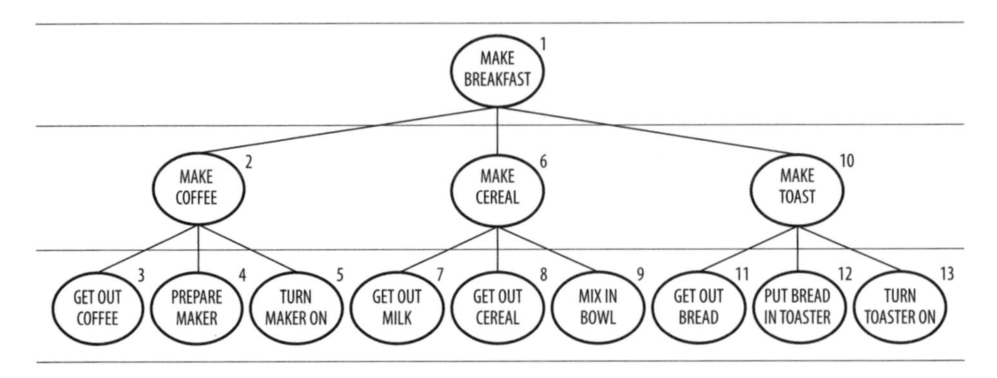

Figure 5.4
An illustration of how activation spreads through the array of goal representations along control lines. Here we artificially take the goals in the bottom row to be basic ones (and omit some obvious goals for simplicity's sake); in reality, many of these goals would control still further goals.

goal representations on and off (Koechlin and Jubault 2006). The whole of the prefrontal area is organized in a gradient fashion, with act representations becoming increasingly abstract as one moves from more caudal (rear) to more rostral (frontal) regions (Badre and D'Esposito 2009; Dixon, Fox, and Christoff 2014).[2] This can be interpreted as the neurobiological instantiation of the different spatial levels of the workspace. It should be noted that this correspondence between the computational model and the neurobiology is not serendipitous, as the model has been developed in part on the basis of neurobiological findings like these.

The same computational system is believed to underlie our ability to construct representations of others' acts. The main evidence for this again comes from neuroimaging studies; we observe a considerable overlap in prefrontal activity patterns when agents perform complex intentional acts and when they interpret them. Unfortunately, the details are seldom worked out, but it is not too difficult to imagine how the system might be run in reverse to yield judgments about the goals underlying an observed act string. When run in this mode, the system might be cued into action by a set of representations corresponding to the basic acts in a string (simple bodily movements). The system might then search for higher-level goals that would be satisfied by groups of these lower-level goals. By recursively applying this process, very abstract goals might eventually be inferred. As with controlling action, dead ends will sometimes be encountered (i.e., no higher-level goal can be identified that corresponds to a grouping), prompting the system to search for an alternative goal at the relevant level, probably with different groupings of basic acts. On this view, both act production and interpretation would be served by many of the same computational operations. If so, then selection for enhanced action planning and control would lead to the selection of abilities that also facilitate interpretation, and vice versa. This would make tool manufacture and use an especially potent driver of upgrades to this computational system. For not only are these skills manually demanding; they are also paradigm cases of socially learned skills, and hence impose demands on the interpretive abilities of apprentice artisans. Building on earlier material, we will shortly argue that these demands had origins deep in the stone-crafting of the Pleistocene.

In fact, the true origins are deeper still. An evolutionary precursor of this computational system is visible in at least some other primates. Premotor cortex F5 in Old World monkeys is thought to be homologous with Broca's area in humans. In macaques, F5 has been shown to contain groups of neurons that are tuned to specific action types such as *tearing, grasping, communicating, eating,* and so on (Rizzolatti, Fadiga, et al. 1996; Gallese, Fadiga, et al. 1996; Ferrari, Gallese, et al. 2003). Moreover, these neurons are supramodal in the sense that they are activated by both visual and auditory information, in addition to being recruited in motor behavior.[3] This is reminiscent of the multimodal character of Broca's area. The latter has been shown to be involved in processing everything from toolmaking to language to music to arithmetic (Fadiga, Craighero, and D'Ausilio 2009). This has led some neuroscientists to describe Broca's area as a "supramodal hierarchical processor," the idea being that it evolved to process hierarchically structured strings *as such* (i.e., irrespective of the sense modality from which the string was originally derived). We think this system was scaled up over the course of hominin evolution, eventually producing our tremendous capacity for representing stimuli hierarchically. We now turn to what we believe was the main driver of such change.

5.6 A Crash Course on Early and Middle Pleistocene Technological Evolution

We have at several places in this book mentioned technological innovations that we think are reliable indicators of hominin cognitive advances.

In this section, we delve into a bit more detail regarding Plio-Pleistocene and middle Pleistocene changes in technological complexity, with a focus on developments requiring the origins and expansion of hierarchical control; of increasing "policy abstractness" of goal representation, to borrow the jargon of neuroscience. We organize this material into three stages. We then unpack the significance of these changes for hominin hierarchical cognition.

Stage 1: Late Pliocene–early lower Pleistocene; Oldowan and developed Oldowan tools. For a long time the production and use of stone tools was thought to have originated around 2 mya with the appearance of *Homo habilis* (the fact that this species was presumed to be the first hominin toolmaker is reflected in its name). Serious doubt has been cast on this view in recent years. Excavations at Lomekwi 3 in Kenya have revealed similar tools dating to as early as 3.3 mya (Harmand, Lewis, et al. 2015). Given what we now know about chimpanzee tool use, that is not surprising. Not only do chimpanzees produce a variety of wooden tools (termite fishing rods, primitive spears); they are also known to transport stone to anvil sites where those stones are used as hammers to crack open nuts (Boesch and Boesch 1983; Carvalho, Cunha, et al. 2008). As a by-product of this process, stone flakes are often produced and can even pile up, leading to chimpanzee flake assemblages (Mercader, Panger, and Boesch 2002; Mercader, Barton, et al. 2007). It may well be, then, that our last common ancestor with chimps and bonobos used stones as hammers, and in doing so produced useable flakes as a by-product. A habitat shift requiring more intense extractive foraging—more resistant plant matter, such as roots; the marrow concealed in the bones and skulls of large animals (Thompson, Carvalho, et al. 2019); raw meat needing tenderizing—could have easily led to the intentional use of these primitive flakes, and then, somewhat later, their intentional production. No significant cognitive or cultural advance would have been required.

So some stone tool use very likely long predated the habilines, with some stone crafting being Pliocene rather than Pleistocene. An untrained eye is unlikely to recognize artifacts crafted before 2 mya—"choppers" or "chopping tools" as they are known—as tools at all. They show no evidence, for example, of further modification; they were neither refined by secondary flaking nor retouched at later times. They appear to be little more than a sharp edge with an area to grip. These implements were probably used immediately after being made and were then discarded. This has led some archaeologists to argue that these tools were not designed at all. To the extent that they have consistent morphological features, that is, to the extent that their shape is predictable, the thought is that this can be explained solely in terms of raw material properties (e.g., fracture properties of the rocks used) together with selection from among a collection of flakes randomly produced by stone percussion. To put it very crudely, bang stones together and then pick up whatever looks most useful. This is a very simple step beyond banging using a stone hammer on a firm base to break open bone, or tenderize a tuber, and in the process occasionally producing sharp shards. The view that the makers of these tools did not have a shape in mind before they began a bout of toolmaking becomes somewhat less plausible from about 2 mya. Beginning around then, we start to see increasing diversification in tool forms: recurring forms of scraping tools, cutting tools, and tools dedicated to percussion. These forms are distinguishable on the basis of their overall size and the properties of their functional end. The term "developed Oldowan" is sometimes used to mark out this change.

Stage 2: Middle/late lower Pleistocene (1.7 mya); Acheulian. The developed Oldowan shades into the early Acheulian. The Acheulian is defined most famously by handaxes, but also by other large cutting tools such as cleavers and picks. By 1.7 mya, these tools had emerged as distinct forms. They typically show bifacial flaking and are elongated relative to earlier tools. By this point, most archaeologists are happy to say that toolmakers were imposing definite form on artifact morphology, though the makers might have been predominantly concerned with just one part of the overall artifact, its cutting edge, together with a secure grip (Kuhn 2020). For these tools could be used with some force. So perhaps these earlier Acheulian tools were only semi-planned. Over the next 0.5 million years or so, more complete control of form emerges. By 1.2 mya, Acheulian tools, and in particular handaxes, had become much more refined. Beginning around this time, knappers began to (primitively) prepare stone cores—pieces of stone from which flakes are struck—so that those flakes would be close to their intended final form when detached from the core. These tools show symmetry along multiple axes and are often very large (some argue so large as to prove difficult to use in practice).[4] Over the next many thousand years, there is progression along all of these design dimensions (bifacial flaking, core preparation, symmetry). This marks the boundary with our third stage.

Stage 3: Middle Pleistocene; Middle Paleolithic technology. Our third stage begins around 800 kya and takes us toward the late Pleistocene. One of the most striking developments during this period concerns the degree of core and platform[5] preparation. Prior to this period, flake removal remained rather simple: flakes were removed in a serial fashion as the core was rotated, and eventually the desired artifact form was brought into existence. This is known as marginal trimming, which causes a rapid reduction in width, but little to no reduction in thickness. That changes around 800 kya. Artifacts begin to show evidence of fully bifacial thinning, which means the removal of flakes across 50% or more of the platform's width. To make long but thin tools, different meta-tools are needed: wood, horn, bone, or soft stone hammers. This technique produces shaped stones that are unmistakable artifacts; there is no doubt that the handaxe in figure 5.5 is a tool. But it is also transformative with respect to the tool's practicality; extensive bifacial thinning produces a high ratio of utility to weight, reducing transportation costs. Compared to their chunkier prototypes, for their length and breadth these tools are much lighter.

Preparatory techniques of this sort had their culmination in the so-called Levallois technique. With this technique, virtually all of the work of knapping was done in advance of removal of the finished flake. Moreover, the core had become the resource, not the tool in embryo. The artifact was removed from the blank (the extensively processed core) with a final hammer stroke with all of its desired functional properties—edge morphology, gripping affordances, resharpening potential—intact. This led to diversification in the blank production process itself, with dedicated blanks of different sizes and shapes. A given blank might also be set up so as to source multiple tools, both at a time and over time, being renewed (i.e., retouched) as needed. The degree of manual precision and foresight shown by these toolmakers is astounding.

Figure 5.5
A late Acheulian handaxe. The Portable Antiquities Scheme, photo by Andrew Richardson, 2004 (CC BY-SA 4.0).

The other, perhaps even more striking development during this period is the appearance of composite tools. Composite tools have many advantages. They are typically mechanically superior, yielding increased force; they are often safer, creating distance between one's hand or body and the target; they are easier to repair. They can be more failsafe, through the addition of redundant elements (for instance, multiple barbs). However, they are also more demanding in at least two ways. First, the manufacture of a composite tool requires sourcing and processing multiple raw materials, so there are increased time and energy costs. Moreover, the manufacture of these tools tends to be more cognitively demanding. It requires skills appropriate to each of these phases of production, and their associated raw materials. In addition to knowing how to produce an appropriate Levallois blank for flake detachment, the artisan must know how to find and shape the appropriate wood for a handle ("haft"), and must master some method for attaching stone to wood. Furthermore, these diverse bodies of knowledge must be brought together in working memory so as to yield a single action plan. This could signal a cognitive advance in its own right—"cognitive fluidity" (Mithen 1996)—but at a minimum it points to upgraded executive capacities, while a more networked population may also have been necessary to support the reliable cultural learning needed to stabilize these skills (Powell, Shennan, et al. 2009; Sterelny 2020).

Since the survival of an actual haft is exceedingly rare, archaeologists have to rely on indirect sources of evidence of hafting. Currently, the earliest evidence for hafting is around 800 kya in the Levant (Wilkins, Schoville, et al. 2012). Here archaeologists have uncovered large collections of flaked artifacts which appear to be too small to be used as hand-held tools. Moreover, the backs of these flakes have been modified in a way consistent with hafting. However, we do not find other aspects of indirect evidence that indicate hafting. For example, characteristic impact fractures and haft wear marks are absent, leading some archaeologists to remain skeptical. By 500 kya, however, we find artifacts in Africa (Namibia) and Eurasia (Germany) which tick more of these evidential boxes, leaving little uncertainty (Wilkins, Schoville, et al. 2012). Still later, at about 250 kya, we find clear evidence for the use of mastics (gum or resins), though it would appear that the first use of mastics was simply to create resin handles for stone artifacts (Mazza, Martini, et al. 2006). Many suspect this date reflects the rate at which the archaeologically known polishes and adhesives decay over time—the oldest possible survival of the evidence, rather than the earliest possible establishment of mastic technology. So mastics and adhesives are at least that old, but perhaps considerably older. But in any case, the manufacture of composite tools hafted with mastics represents an unquestionable advance in both cognitive and cultural sophistication, given the demands of this technology on planning, action organization, and knowledge. For mastic production itself is a complex behavioral affair (figure 5.6). For example, many adhesives can be made only through the quite precise application of heat.

This takes us to the brink of the late Pleistocene. This saw a whole host of additional technological innovations, including high-velocity projectile weapons. As we shall show in the chapter 7, these had quite profound social implications. They may well have been of importance to our species' survival in the last ice age, as that seems to have been even more climatically variable than earlier ones (Martrat, Grimalt, et al. 2007; Loulergue, Schilt, et al. 2008). We suspect that these developments primarily, or even entirely, reflect cultural changes in our line, rather than the addition of new cognitive equipment. Some of those developments will be reviewed in the next chapter; they are not directly relevant to our analysis here. In our view, these middle Pleistocene technological developments depended on the expansion of hierarchical control and its recognition, and hence

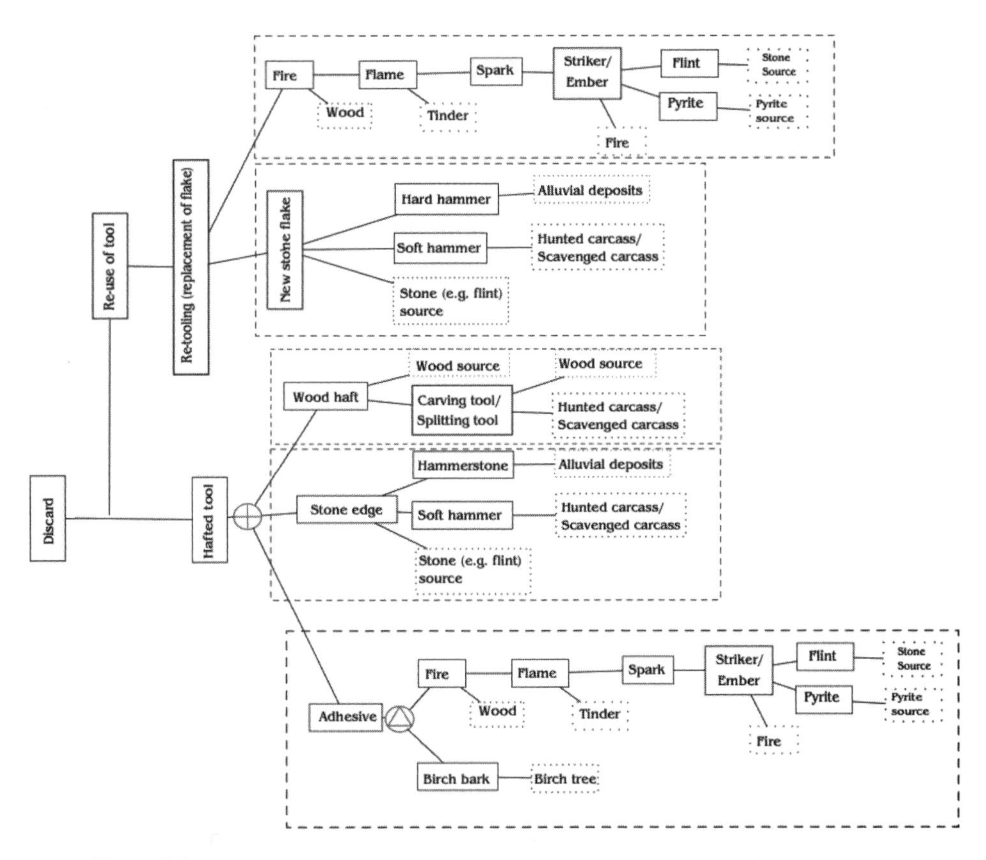

Figure 5.6
Sykes's diagram illustrating a possible *chaîne opératoire* for composite tool manufacture using birch bark pitch. The diagram simultaneously represents three courses of action: (i) the *de novo* creation of a hafted stone tool; (ii) the retooling of an existing hafted tool with a new flake; and (iii) disposal of the tool. The Hafted tool node controls three subgoals at the next level down, Adhesive, Stone edge, and Wood haft. And each of these, in turn, controls additional subgoals. Likewise, the Re-use of tool node controls two subgoals, New stone flake and Fire, which in turn control other less abstract goals. Material within a dotted box corresponds to possible goals (e.g., the toolmaker may not have to source stone if some is ready at hand). (The reader will notice that this diagram makes use of some different conventions for representing the structure of an action plan than we have used in this book; but it is easy enough to see how the information in the diagram might be reorganized to fit our format.) Figure reproduced from Sykes (2015) with permission of the author.

selected for the computational capacities that made it possible for hominins to fluently use hierarchically structured sentences.

5.7 The Evolution of Human Hierarchical Cognition

We take this story of technological change in our line to suggest the following evolutionary scenario. The comparative data suggest that the last common ancestor of Old World monkeys and apes possessed a computational system specializing in action organization and recognition. This system became more complex in the ape line as hominoid foraging techniques became more complex. Great apes in general are good at making and using tools, and we take that to show that the first great apes already possessed a significantly upgraded version of this system. We suspect great apes employ and recognize at least some higher-level goal representations. If so, then they have some capacity for constructing hierarchical representations of act strings. The same would have been true of the very earliest hominins. This system continued to evolve in our line, undergoing profound change as hominin technological complexity increased. Oldowan tools were made of stone, and so we know about them. But their makers probably did not need cognitive mechanisms of action representation that were much more sophisticated than those of great apes; perhaps indeed they were no more sophisticated. However, that is not true of Acheulian technology. As mentioned above, after 1.7 mya we begin to find handaxes that could be made only with extensive planning and control. The manufacture of a thin, elongated, highly symmetrical handaxe would have almost certainly required a complex action plan, where one or more of the components of the plan could be flexibly extended or modified as needed, in the face of mistakes or unanticipated outcomes. These demands only intensify with the transition to middle Pleistocene technique and technology. Skilled use of the Levallois technique, or the production of a composite tool using a carefully prepared mastic, depends on capacities approaching, or on par with, those of fully modern humans.

In keeping with the gradualist, neo-Darwinian character of our approach to language evolution, we would like to know how human hierarchical cognition might have evolved via a series of small, incremental changes. Fortunately, this is relatively easy to see in the case of the computational system set out in section 5.5. Below are four ways in which the system might have been gradually upgraded, and there are doubtless others:

- *Activity maintenance.* The control of extended action sequences depends on maintaining goal activation. A higher-level goal is active while the lower-level goals it controls are carried out. The time a higher-level goal can be kept active sets a limit on the complexity of the sequence of lower-level goals it can control. Premature decay of a higher-level goal would effectively result in forgetting what one is doing: losing one's thread, as indeed we sometimes do. Hence the system can be made more powerful by increasing the time over which goals can be kept active (and/or by increasing its stability; this applies below as well).
- *Length of planning.* The number of goal representations that can be held within a given row of the workspace might increase over time. As this number increases, action plans could become longer. They would stretch further into the future, not just in time but in the number of elements in the action plan.
- *Depth of planning.* Length of planning is a horizontal change. The control system might expand in depth, too. That is, the number of levels that an action plan can contain might increase. This would result in action plans having deeper hierarchical structure.
- *Rapid transitioning.* Action is controlled by switching on and off act representations within the action plan. The rate at which this happens determines the speed and fluidity of the actions being performed. One way the system might be made more powerful is by increasing this rate. One striking feature of skilled action is this increase in speed and fluidity. It can seem completely automatic and reflex-like until the unexpected happens; then, skilled agents show that fluent execution is not reflex-like by responding in ways highly sensitive to context. We know humans develop these highly expert capacities through practice (but not only practice). There does not seem to be an equivalent of expertise in great ape behavior. We show these kinds of skills in language too, when we successfully interpret conversational surprises and misfires, still quite quickly and still quite reliably. Most jokes depend on conversational surprise not derailing successful interpretation.

The thought, then, is this. Technological evolution in our line drove the evolution of a system for the hierarchical control of complex, extended, and precise action plans, and in social learning for the parsing of the complex actions of other agents. This system was an elaboration of capacities already possessed by the ancestors of great apes and hominins: it did not come from nowhere. Once this was in place, our ancestors had all they needed, cognitively speaking, to make use of hierarchical structure in language. They would have been able to represent sentences as having nested part-whole structure. This would have made possible the eventual evolution of conventions that gave a role to such structure in encoding meaning.[6] They also had all they needed to put to use hierarchical structure below the sentence level, yielding morphological principles, and perhaps above this level as well. From a computational perspective, these ways of making use of structure in language are not fundamentally different. As with simple syntax, we think many of these conventions arose out of default patterns that carried information about speakers' intentions. Those patterns were then stabilized and refined through sender-receiver coadaptation. Especially early on, we suspect that the role of comprehension (in Dennett's sense) in shaping these conventions was minimal.

One might doubt that a system designed for planning and recognizing action could have been co-opted for use in linguistic communication. It is true that we are accustomed to thinking of action and linguistic communication as fundamentally distinct. But they are not really. Language is a special case of complex intentional action. That is one of the insights of the pragmatic approaches to language we have already discussed. It is no accident that one of the foundational texts in pragmatics is called *How to Do Things with Words* (Austin 1962). But more importantly, the neuroscience is consistent with this assumption. For as was noted earlier, Broca's area and other nearby regions are paradigmatically multimodal. It seems that the areas that carry out the relevant computations simply do not "care" whether it is toolmaking or language or any other complex intentional activity that they control or identify. This is not to say, however, that language has not exerted an evolutionary influence on the structure of this computational system. We are certainly open to that possibility. Indeed, we would be most surprised if the evolution of language made no difference to these control and recognition mechanisms. For example, the demands of conversation may have selected for increased speed. The fluency of skilled action in expertise might thus in part derive from selection for fluent communication. In addition, once language was primarily spoken, there may well have been selection for processing much more fine-grained stimuli. Once anything approximating language has evolved, there are an enormous number of functionally distinct options that are available as potential acts, and these differ from one another only in physically subtle ways. In both executing these and recognizing them as distinct, precise control and fine-grained analysis are essential, and we think this is likely to help explain the origins of phonemes. But these changes would have been changes laid over top of an already quite sophisticated system, one that originally evolved, we think, in the service of technological skill and the learning of such skill from others.

Before closing this chapter, there are a few features of our account of hierarchical structure that are worth highlighting. First, on our account, the very same (linguistic, action) string can have hierarchical structure, or have a particular kind of hierarchical structure, for one agent but not for another agent. Hierarchical structure is thus agent-relative on our account. Indeed, it is relative to an agent at a time, as the same agent might represent a string differently over time. A string might be represented differently depending on the context and its stakes. That would be natural if cognitive costs/demands scale with the complexity of hierarchical representation.

Such a view of hierarchical structure contrasts with the standard linguistics conception, which encourages us to think about hierarchical structure as if it were an invariant property of a string itself. On that conception, a string cannot be structured differently for two individuals who are regarded as speaking the same language, and hence as "knowing" the same formal grammar. In our view, thinking about hierarchical structure as "given" in this way masks a serious explanatory challenge. In order for linguistic conventions involving hierarchical structure to arise and stabilize, there must be some correspondence between how senders and receivers represent strings. However, the correspondence need not be perfect; they need not agree on every detail of how the sentence is hierarchically structured. But for a convention involving hierarchical structure to do real work for a speaker and hearer in a given case, some correspondence is necessary. A sender-receiver perspective on the evolution of such conventions enables us to see how this might go.

Second, and relatedly, our account makes room for a disconnect between the hierarchical structure ascribed to a string by a formal grammar and the hierarchical structure that is ascribed to the string by actual speakers/actors. As Frank, Bod, and Christiansen (2012) point out, analyzing sentences hierarchically may prove theoretically useful or convenient, but whether (and how) agents actually make use of hierarchical structure in cognition is a further issue. So there are (at least) two separate sets of facts here. There is the structure specified by a formal grammar, and there is the structure imposed by agents' representations. These two forms of structure might coincide, or might come apart in quite significant ways. In line with our earlier remarks, we suspect the latter is more often the case.

Third, our account stakes out a helpful middle ground between two theoretical extremes in the literature. One of the main reasons Frank and colleagues are skeptical of the role of hierarchical structure in language processing is that they think other animals, including great apes, lack the capacity for hierarchical cognition.[7] They thus see their deflationary view as more attractive on evolutionary grounds; there is no glaring gap that needs filling between our capacities in this area and those of other great apes. Perhaps the most vociferous criticism of Frank and colleagues' line is from Berwick and Chomsky (2016). They tell us, "While it is indeed true that the evolutionary story becomes simpler if the only thing all animals can do is process sequentially ordered items, this position has a problem. It's wrong. Hierarchical representations are omnipresent in human language syntax" (p. 115). Both sides agree that hierarchical cognition is off-limits to other animals. For Frank and colleagues, this is a reason to (quite implausibly in our view) deny that hierarchical cognition is involved in human language processing, or to claim that it plays at best a minimal role. For Berwick and Chomsky, it is the key difference maker; it is what explains why humans, and humans alone, have language. We think both sides are wrong. As we said earlier, we think it is quite likely that other great apes at least have some capacity for hierarchical cognition. This is made plausible by the sophistication of great ape action and (arguably) their social learning. It is just that this capacity is much more rudimentary in them. Over the course of human evolution this capacity was transformed as a result of selection for enhanced technological capacities. A smooth progression takes us from the great ape baseline to humans' current abilities to routinely use hierarchical structure in language.

Finally, it is worth briefly contrasting our account with one that has received a good deal of attention from archaeologists. Stanley Ambrose (2010) has argued that humans' syntactic capacities evolved in the service of composite tool manufacture. His guiding thought is simple: to make a composite tool, one must be able to think about combining, and possibly recombining, various parts into a functional whole. Moreover, these parts must often be manufactured and combined in a specific order. The suggestion is that these same capacities were then co-opted for use in the linguistic domain, allowing us to combine and recombine linguistic units. Not surprisingly, we think Ambrose is right to emphasize the cognitive demands of manufacturing composite tools in attempting to explain the origins of human syntactic capacities. The problem is that Ambrose ignores other technological developments that are equally important to the story. He does this in two ways. First, he treats earlier forms of technological change as more or less irrelevant. That is a mistake. For even though Acheulian stone tools were not multicomponent tools, they were difficult to make, and selection for the abilities needed to make these tools built the capacities to use syntactically structured signaling systems. Second, while it is natural enough to draw an analogy between combining artifact components and combining words, the analogy is probably superficial. The cognitively critical issue is the nature of the plans agents have to form and execute (or recognize, as interpreters), not the nature of the products of the plan. Hence, in our view, the critical feature of both cases is the complexity of action plans and their interpretation. The manufacture of a thin, long blade using the Levallois technique is surely highly complex. So while on the right track, Ambrose's focus is much too narrow.

A lot of ground has been covered in this chapter. We began by considering the syntactic capacities of some of our closest relatives. We argued that great apes appear to be cognitively equipped, or very nearly so, to use simple forms of syntax. This fact (if it is one) goes a long way toward demystifying the evolutionary origins of simple syntax in our own line. In our view, the primary explanatory challenge we face is humans' prodigious ability to entertain and make use of hierarchically structured representations. This, we argued, is best explained by selection for enhanced technological capacities operating over the last several million years. Human hierarchical cognition was built incrementally over this period, and was co-opted for use in language later.

6 The Firelight Niche: From Sign to Speech

6.1 A Gesture-Speech Transition

So far, we have defended a version of the gesture-led theory of language origins. Our version of this theory acknowledges that communication in our line was very likely multimodal from its inception, incorporating at least some vocal elements. This multimodal hypothesis we take to be all the more plausible in light of more recent evidence demonstrating greater top-down control of vocalization in other great apes than was once supposed. But we still claim that it was gestural communication that did a lion's share of the communicative work early on, and that gesture remained dominant for some time. To briefly recap the arguments of sections 2.3 and 2.4, a gesture-visual channel leads much more seamlessly to a structured rather than a holistic sign system. Moreover, while there is now evidence of somewhat greater top-down control of vocalization in great apes, that control is much more complete in the manual domain; great apes have full or near full top-down control of their hand movements. These facts were almost certainly true of our last common ancestor with the other great apes. Further, while Kanzi and company show iconicity was not essential to secure uptake, the gestural-visual channel has a greater potential for iconic communication, and that would have made it possible for communication to expand in our lineage with only a marginal increase in cognitive sophistication compared to other apes. With that expansion came enhanced cooperation and cultural learning, in turn creating a coevolutionary dynamic leading to selection for a further expansion of communicative capacities. At about the same time, our ancestors were undoubtedly experiencing strong selective pressure for enhanced manual capacities subserving extractive foraging and tool manufacture and use. Many, perhaps most, of these techniques had to be socially learned, and so hominins would have been being selected for greater attention to the intentional activities of others' arms and hands. These changes to motor control, planning, and attention built the capacities needed for an expansion of gestural communication.

There are still other advantages, but the above is already enough in our view to provide substantial support for a gesture-led theory. However, any theory of this sort faces a major explanatory challenge. For language as we now know it is predominantly vocal-auditory. The challenge, then, is to provide an account of the transition from a primarily gestural to a primarily vocal mode of communication. Good "scenario-building" (Sterelny 2012a) in turn requires that this account be embedded in a larger account of human biological, social, and cognitive evolution for which there is independent empirical support. It is not enough to simply cite advantages of vocal communication over gestural communication. For that does not specify a pathway or a context.

If a gesture-led theory is right, there has been a profound change in the importance and relative role of gesture and voice. But gesture remains important. Very rarely do we communicate face to face with little gesture; indeed we often gesture even when we are not talking face to face, for example while on the phone. This is especially clear once we recognize that gesturing is not limited to the hands; we routinely gesture with our posture, eyes, eyebrows, noses, mouths, and other bodily parts. Such displays are not epiphenomenal. Gesture works together with speech in some obvious ways—uttering "it's him," while openly gazing in a certain direction—but the effects of gesture go well beyond these cases. This has been persuasively demonstrated by Goldin-Meadow and others (see, e.g., Goldin-Meadow, Wein, and Chang 1992; McNeill, Cassell, and McCullough 1994; Thompson and Massaro 1994; Alibali, Flevares, and Goldin-Meadow 1997; Kelly and Church 1997; Goldin-Meadow and Sandhofer 1999). To give just one interesting example: Goldin-Meadow and her team have extensively studied the role of "gesture-speech mismatch" in human communication. As the name suggests, these are cases in which an individual's gestures and speech embody conflicting information. These cases are methodologically informative, for they feature gestures that carry different information from the accompanying speech, and hence it is easier to confirm that the audience is actually picking up on the gestural information. When, for example, gesture just adds emphasis or redundancy, it is much harder to show that the gestural contribution has been salient to audience uptake. Goldin-Meadow (1999) reports one memorable case involving a six-year-old child's judgments regarding the numerosity of two rows of checkers. Initially, the checkers had been lined up in rows of equivalent sizes, and the child correctly verbally reported that the number of checkers in each row was equal. But after one of the rows was spread out, the child reported the spread-out row was now "different because [the experimenter] moved them." At the same time, however, Goldin-Meadow describes how the child gestured in a way that suggested he still perceived a one-to-one correspondence between the two rows (he gestured back and forth between the checkers in the two rows). An untrained adult asked to assess the child's reasoning spontaneously attributed to the child an understanding that the two rows of checkers could still be placed in a one-to-one correspondence.[1]

There appear to be various noncommunicative benefits to gesturing as well—for example, gesturing can help us retrieve information from memory (Rauscher, Krauss, and Chen 1996) and reduce cognitive load (Alibali and DiRusso 1999; Kirsh 1995)—but the important idea for present purposes is that gesture continues to be significant in everyday conversation.[2] However, when gesture is coupled with speech, it is not structured in the way speech is: we observe none of the morphological or syntactic structure of natural language sentences in the gestures accompanying ordinary speech. As evidenced by sign languages, and even more strikingly by cases of homesign, gesture remains capable of functioning linguistically when it must. We take this this to be a point in favor of gesture-first theories, for they can say that this capacity reflects a stage in the evolution of language in which gesture was the primary means of linguistic or protolinguistic communication. In contrast, there is no obvious explanation for this capacity on vocal-first theories of language. For notice that there is no general capacity to spontaneously invent or learn to use any suitable physical medium for language. Children have to be explicitly taught the skills of literacy. They do not routinely, spontaneously pick them up, even in cultures where they extensively co-read with illustrated children's books. Still less do they spontaneously invent language-as-writing, as we see with homesign. They have an untutored capacity to co-opt vocalization or sign for language-like communication, and only vocalization or sign. But when spoken language is available, gesture is—barring very special circumstances—relegated to an auxiliary role in communication. So we do not have to explain why gesture "went away," for it did not, but we do have to explain how it gradually came to take a backseat to talking.

We begin with morphology and physiology, before outlining our view of the selective contexts and pathways though which a predominantly gestural system became primarily vocal. What changed in hominin bodies to make a central role for speech possible, and when did those changes take place? As we will see in section 6.6, the morphological and physiological changes that make modern forms of speech possible do leave traces in hominin fossil anatomy. But it would be rash to treat those morphological changes as by themselves dating the origin of a primarily vocal form of (proto-)language. Early vocally dominant hominins might have made do with a simpler and cruder set of vocal contrasts, especially if they still used a relatively sparse lexicon. Moreover, song requires similar adaptations as speech, so these changes might reflect an increased importance of musicality.

6.2 The Anatomy of Speech

Our capacity for speech depends in part on unique features of human cognition. In particular, we cannot learn the vocabulary of a vocal language without vocal imitation. Some birds are mimics, but vocal mimicry is not found amongst other primates.[3] This conspicuous difference makes its evolution in our line a serious puzzle. In other respects, our capacity to talk is dependent on systems that exist in other mammals, though modified for the special demands of speech. So, as Tecumseh Fitch has emphasized (2000; 2010), with the exception of this capacity for vocal imitation, it is much easier to understand the evolution of speech than it is language. For while there is no clear homologue of human language in the animal world, all mammals produce and perceive vocalization in essentially the same way. On the production side: air is exhaled from the lungs, causing vibration of the vocal chords in the larynx. These vibrations then pass through the vocal tract—a series of cavities—where they are modified in various ways by the shape and position of vocal articulators. The result of these modifications are *formants*, dense bands of acoustic energy in the resultant air vibrations. When these vibrations reach others, they enter the ear canal where they are transduced into neural signals via the activity of hair cells in the inner ear.

Looking at it from this perspective, we can readily identify the unique features of the human vocal-auditory system. The most obvious difference is the resting position of our larynx, which sits much lower in the throat than in other mammals. It makes its initial descent between the ages of three and four, and descends even further in males during puberty. This has the effect of lengthening the vocal tract, making possible a wider range of sounds. When this is coupled with the increased innervation of vocal effectors—of the tongue and lips—a wider range of formants can be sounded. Humans also show enhanced breath control relative to other mammals, including great apes. Anatomically, this is reflected in increased innervation of our diaphragm and intercostal muscles (i.e., those between the ribs). Such control is critical to speech, with its precise demands on the timing and duration of inhalations and—often sustained—exhalations. In addition, the laryngeal air sacs characteristic of mammals were lost at some point in human evolution. The function of these air sacs is still a matter of debate, but most theorists believe they play a role in amplifying and/or deepening vocalization. The loss of these air sacs increased the range of distinct speech sounds we could produce (de Boer 2009).

So modern speech relies on a suite of interacting changes to our vocal tract and respiratory system. On the receiver end, our auditory system changed in ways that boost our sensitivity to pitch within certain frequency bands; specifically, compared to other primates we show heightened sensitivity to frequencies within the 200–600 Hz range (Martínez, Rosa, et al. 2004). But speech is also associated with some "higher" neurocognitive changes. In particular, as we have already mentioned, humans show much greater top-down control over vocalization. Human speech is the product of two interacting control systems in the brain (Ploog 2002). One of these is shared with a wide range of taxa. This system is built out of limbic-system components and underlies emotionally laden vocalization. The other system, which is unique to primates, connects neocortical areas with vocal articulators via the hypoglossal nucleus (located in the medulla). This neocortical pathway is more highly developed (i.e., is made of longer and thicker fiber tracts) in humans than it is in any other primate. In primates, damage to the neocortical pathway leaves spontaneous vocalization intact, while in humans, it completely disrupts normal speech. This reflects the fact that language is under voluntary control and yet, even so, often serves to express emotion (as in vitriolic attacks or cathartic rants). Our superior control over the vocal chords, tongue, and mouth is the result of the greater development of this neocortical pathway in our lineage.

We have neither the capacity nor the intent to do full justice here to all the intricacies of the evolution of speech. The take-home message is that our ability to speak and listen depends on a suite of morphological and neurological changes. These presumably developed over significant time and from sustained selection pressures. In the rest of this chapter we outline some plausible drivers and pathways through which primarily gestural communication became primarily vocal. Our central idea links this transition to another revolution in human life, the domestication of fire, and its consequences for nutrition, social organization, and time management. We begin with an apparent triviality: the more our ancestors used their mouths for utilitarian purposes, the less mouths were available to broadcast sound. That constraint is less trivial than it seems.

6.3 Fire, Cooking, and Freeing the Mouth

It is tempting to think that prior to the evolution of speech our mouths were more or less unoccupied, free to be recruited into communicative service. Reflection on the amount of time other great apes spend chewing per day tells a very different story. It is estimated that chimpanzees and orangutans spend around 7 hours per day feeding (Ghiglieri 1984; Wrangham 1977; Galdikas 1988), while gorillas spend some 8.8 hours per day on this activity (Lehmann, Korstjens, et al. 2008; Ilambu 2001). These numbers are all the more impressive given that great apes rise with the sun and go to bed at sunset, implying an effective day length of about 12 hours in low-latitude regions. In contrast, modern humans spend a little over an hour a day engaged in eating and drinking (Bureau of Labor Statistics 2020). Great apes are big-brained and big-bodied and hence have high caloric needs. The enormous amount of time they spend feeding each day reflects the demands of eating unprocessed food. Such food not only demands longer, more vigorous chewing; it is also harder to digest. For even after prolonged chewing, the food is still not as broken down as processed foods. In apes, a diet composed entirely or mostly of unprocessed food imposes strong selection pressure for a powerful craniofacial anatomy and large teeth (Wrangham 2017).

Richard Wrangham cites these contrasts between humans and great apes in feeding time in developing his hypothesis that humans are now, and have long been, dependent on processed foods. Indeed, in his view, humans are obligate consumers of *cooked* foods. This is very plausible.[4] But he extends this back to *erectus*, arguing that the erectines regularly consumed cooked foods. As evidence, he cites erectines' more gracile craniofacial anatomy, less powerful jaws, and smaller teeth, together with other anatomical changes, for example a shorter gut and ribcage. Fonseca-Azevedo and Herculano-Houzel (2012) tested this idea. Based on brain and body size, they estimated the amount of time various hominin species would have had to spend feeding, given a diet of only raw food. They found that, whereas australopithecine and habiline feeding times would have been on par with those of *Pan* and *Pongo* (about 7 hours per day), erectines would have had to spend 9 hours per day feeding, and later hominins 10. As these authors explain, it is a struggle for other apes to feed for 8 hours per day and still meet the other demands imposed by their lifeways (e.g., grooming, rest). For example, in periods of heavy rainfall, gorillas will halt eating temporarily, but if the rain continues for more than 2 hours, they resume feeding and make up for lost time by resting less on those days (Watts 1988). This suggests that it would have been all but impossible for erectines to chew for 9 hours a day, given the much greater complexity of their social and economic worlds. These time costs might be reduced by eating more calorie-dense foods—specifically, meat. However, it is very unlikely that meat alone could reduce feeding times to feasible levels, given the unreliability of this resource, and the fact that it takes a large time and energy investment to obtain (Wrangham 2017).

Fonseca-Azevedo and Herculano-Houzel take these estimates to support Wrangham's hypothesis. But as Wrangham himself points out, processing need not involve cooking: food can be mashed or ground up, thereby making it easier to both chew and digest (Wrangham 2017). Unfortunately, Fonseca-Azevedo and Herculano-Houzel did not consider the option of processing without cooking, perhaps supplementing such a diet with fatty animal products like marrow and brain, and whether such a diet might have adequately driven down feeding times. The plausibility of this alternative explanation depends on the extent to which cooking is superior to other forms of processing in generating caloric gains. This has been investigated experimentally in mice by Carmody, Weintraub, and Wrangham (2011). They compared the effects on net body mass of a diet of (i) raw, unpounded food; (ii) raw, pounded food; (iii) cooked, unpounded food; and (iv) cooked, pounded food. They did so independently for both a starch-rich tuber diet and a meat diet. The team found that cooked tubers were roughly four times as energy-efficient as raw, unprocessed tubers, and that raw, processed tubers were roughly twice as energy efficient as raw, unprocessed tubers. With meat, cooking mattered but other ways of processing did not: eating cooked meat was found to be roughly twice as efficient as eating it raw, whether processed or unprocessed.[5] Cooking probably reduces mastication and digestion costs, and, with meat, reduces immune costs, as cooking serves to kill harmful pathogens.

Perhaps unexpectedly, this excursion into the prehistory of cooking is relevant to the evolution of speech. First, a morphology ideal for chewing and swallowing tough food is not ideal for controlled articulate speech, and vice versa. Making food softer and easier to fragment eases constraints on mouth, jaw, and vocal tract imposed by the need to chew through tough food. Once food is processed, there is less of a tradeoff between a system optimized for chewing and swallowing and one optimized for speaking, though there is still some tradeoff. In particular, the low position of our larynx is notorious for the increased risk of choking. As emphasized by Lieberman (2012), this cost should not be underestimated. As of 2017, choking was the fourth leading cause of unintentional death by injury in the United States.[6] This remains true despite widespread cultural awareness of choking risks, and most people's familiarity with the Heimlich maneuver. The risks to children are especially acute. It seems quite likely that cooked food poses a considerably lower risk of choking than does raw food, as it more easily fragments into smaller pieces. Perhaps that is even true of raw but pounded food. In any case, the longer the feeding time, the longer the exposure to the risk of choking, and cooking certainly reduces feeding time very significantly. Second, and in our view most importantly, cooking implies some control of fire, and we shall argue in section 6.5 that fire changes the social environment in ways that favor greater reliance on talking. Third and most obviously, if cooking dramatically reduces feeding time, it frees the mouth and its associated equipment for other activities. These might well include laughter, song, and even the use of the teeth to grip and hold soft materials as well as speech or its prequels.

How much time was released, and how critical was that release of time? That is hard to estimate. Time release depends both on the kinds of food processed and the type of processing: cooking versus pounding and/or grinding. Let's begin with the supposition that early hominins mostly ate plants, as do many ethnographically known foragers. If the energy ratios Carmody et al. report approximately hold for those hominins, then pounding and grinding food before eating would cut feeding time roughly in half (raw, processed plant food is twice as energy-efficient as raw, unprocessed plant food). Cooking would cut the time by three-fourths. So processing plants without cooking suggests 4.5 hours of feeding time per day for erectines and 5 for later *Homo*. (Recall the estimates of Fonseca-Azevedo and Herculano-Houzel for eating times given a diet of raw, unprocessed foods: 9 hours per day spent feeding for erectines; 10 hours per day for later, bigger-brained hominins.) That would still be a lot of time spent chewing. However, if these hominins routinely cooked their food, then erectines and more recent hominins would feed for less than 3 hours per day. If meat were a substantial proportion of their diet, these estimates fall again, toward a figure that resembles modern feeding times. The bottom line: any systematic processing saves significant time (given the huge mechanical advantage of tools over teeth). Cooking saves even more time, especially for those hominins eating largely plant-based diets.

How significant for the evolution of speech was the availability of this extra time? Again, that is difficult to estimate. If hominin communication was brief and staccato, long feeding times would not inhibit the use of the vocal channel. If communication involved extended bouts, then the time freed looks more important. This issue becomes central in the next section, where we focus on the role of laughter, song, and talk in social cohesion. These are not brief and staccato uses of the mouth, and we shall suggest they have ancient roots, with the erectines. If and to the extent that such communication was central to social bonding and defusing conflict, the relaxation of the feeding-time constraint mattered.

6.4 Laughter and Song

In chapter 3, we argued that the cognitive capacities that made elaborated gesture possible evolved to control skilled action. In a somewhat similar way, we shall argue that the capacities that made speech possible—precise top-down control of vocalization, and the capacities to attend, recognize, and match others' vocalizations—primarily evolved for song, or song's evolutionary antecedents. They were then co-opted in the evolutionary transformation of multimodal communication, as the vocal component became more prominent and its role changed. In developing this view, we will use for our own purposes recent work by Robin Dunbar (Dunbar 2014;

2017). Like us, Dunbar sees the evolution of laughter and song as building a platform for the evolution of speech, with his account couched within his social brain framework. On this hypothesis, encephalization in our line has mainly been driven by the complexities inherent in living in large primate social groups, and Dunbar and his colleagues have built an equation linking neocortical volume to group size. Dunbar uses this equation to estimate the effective group size of different hominin species. In turn, he estimates the amount of time each hominin species would have had to spend grooming one another to maintain social cohesion, making use of a general equation linking these two variables in monkeys and apes. According to his estimates, australopithecines would have had to spend between 1 and 2 hours per day physically grooming one another. He suggests that they could not afford much more time than this, given the time demands of other activities (feeding, moving, resting). Here we again meet the issue of hominin time budgets. The increase in group size in *Homo* predicated by the increase in neocortical volume implies a corresponding increase in the time demands of grooming.

On Dunbar's logic, by the time we reach *heidelbergensis* and anatomically modern humans (AMHs), and quite possibly well before then, these demands would have been impossible to meet. Specifically, he calculates a time budget deficit of 2.3 hours per day for *heidelbergensis*, and 3.1 hours per day for AMHs. Dunbar suggests that language absorbed much of the work of grooming, developing this view in his 1996 book *Grooming, Gossip and the Evolution of Language*. Grooming is one-on-one. Gossip is more time-efficient, as it is a many-many interaction. Dunbar now suggests that before we spoke, we sang, and that before we sang, we laughed, with these activities likewise serving to socially bond us. We are very skeptical of the supposed equations linking group size to encephalization. They do not work very well for great apes, and we suspect that foraging demands rather than predation threat primarily determined group size. But we do think social complexity went up, and with it social stress. For hominins lived lives that were increasingly interdependent, yet with ever-present tensions about the division of resources. Their lives were a difficult balance of dependence and competition (Kelly 2013). So Dunbar's insights about time budgets and conflict management can be detached from his views about group size. We agree that with more recent hominins, conflict management was more challenging, and that it is very plausible that laughter and song played important roles in supporting social cohesion. We have a different account of the importance of gossip, but that is an issue for the next chapter.

It is indeed possible that the breath control on which laughter now depends evolved for singing (or speech). But we find the suggestion that laughter was an early and important mechanism of social bonding plausible. For as Dunbar points out, we share laughter with the other great apes. However, our laughter differs from theirs in an important way, for our laughter is associated exclusively with exhaling. In contrast, great apes repeatedly inhale and exhale while laughing; not unlike panting (Provine 2001). This difference is due to differences in breath control (MacLarnon and Hewitt 1999). Dunbar suggests that the release of endorphins associated with human laughter is likely a response to the bodily stress we experience during laughter (specifically, stress to the diaphragm and the chest muscles, coupled with temporary oxygen deprivation) (Dunbar, Baron, et al., 2012). In triggering the release of endorphins, the neuroendocrine effects of laughter resemble those of grooming, and so in many contexts laughter would have the same effect as grooming. This is important, as laughter is more efficient. Only the individual who is being groomed tends to produce endorphins, whereas laughter produces endorphins in all who laugh. According to Dezecache and Dunbar (2012), the average size of laughter groups in ecologically valid settings is three, making laughter three times as efficient as grooming, even if a grooming bout takes the same time as a joke (practical rather than verbal).

If laughter bonds more efficiently than grooming, singing is more efficient still. The idea that song was a precursor to human speech has a long history—Darwin himself proposed such a view.[7] But it has typically been developed in the context of sexual selection, with males singing to impress females. Dunbar's account is distinctive in suggesting that singing functioned primarily to promote group cohesion. To the best of our knowledge, this possibility was first raised by Richman (1993). As Dunbar explains, singing also triggers the release of endorphins. But in addition, when performed in a group context, it tends to produce feelings of belongingness (Pearce, Launay, and Dunbar 2015; Weinstein, Launay, et al. 2016). Moreover, as the ethnographic record shows, people frequently sing in groups much larger than three—indeed, an entire band might sing together. Our view differs from both Dunbar and Darwin in that we think that while singing is a precursor to speech, it is not a precursor to language. Rather, singing helps explain the transition from gesture dominance to speech dominance.

The idea that (wordless) singing is a precursor to speech has important explanatory advantages. First, as pointed out by more or less everybody, singing and speech depend on many of the same features of our vocal tract and breath control. Hence, if there is reason to suspect singing evolved earlier in the hominin lineage, its evolution explains, or partially explains, the evolution of those features in our line. Second, in sharp contrast to laughing, singing is a prototypically voluntary activity. When we sing, we in general intend to do so, and we are always exercising some degree of top-down control over our voices. Thus, the demands of singing can help to explain the elaboration of the neocortical pathway involved in human speech. Third, to produce the feelings of belongingness to a group, it is important that individuals coordinate their singing. It must be experienced as a joint activity. While this might take the form of different parties singing different parts (imagine a kind of call-and-response theme), the simplest format would be for people to simply sing in unison. But that requires the ability to match others' vocalizations—in other words, vocal imitation, and, more generally, intentional listening to others.

Hominins who can sing, albeit wordlessly, and who can use a reasonably rich, largely gestural protolanguage, though with some vocal elements, are poised to make a transition to a primarily vocal system, if selection favors such a transition. To the extent that earlier systems are multimodal, they are already primed with the realization that sound can carry or modulate meaning. They have voluntary control over sound production that suffices for reasonably precise and extended sequences. They have the cognitive tools to recognize, respond to, and learn to reproduce others' sound sequences. It is true that such a shift surrenders the advantages of iconicity. But gestures in regular use become stylized and conventionalized; even chimp gestures do. Once gestural protolanguages are well established, their stability does not depend on iconicity, though presumably it remained and remains an advantage in coining new signs.

6.5 The Firelight Niche[8]

Let's recap the argument of the chapter so far. We have argued that hominin communicative capacities initially expanded largely through increasingly elaborate use of gesture. If that is right, at some stage in our evolutionary history there was a transition to a primarily vocal mode. That transition was not trivial. Human mouths, vocal tracts, and the neural machinery of vocal control are all quite different from those of great apes. Those differences in part flow from the fact that great apes do not speak and we do, though they are connected to diet too. We discussed those changes in sections 6.2 and 6.3. In the last section, we argued that the critical preconditions for speech as detailed in 6.2 and 6.3 largely evolved for song and/or its precursors. That was possible only once time constraints imposed by feeding had eased. In developing our view of the role of song, we borrowed a line of reasoning from Robin Dunbar. As hominin life became more complex, maintaining social cohesion and keeping conflict within bounds imposed greater demands, including greater time demands. Hominins retained the great ape social technology of grooming—of intimate one-on-one contact. But Dunbar argues, and we agree, that it was supplemented by more inclusive social technologies, laughter and proto-song. In the last two sections of this chapter, we add the final elements to our account of this transition. In this section, we outline our account of the selective features responsible, and in section 6.6 we anchor the whole account in the paleoanthropological record.

Talking offers some obvious advantages over signing. It allows us to freely use our hands for other tasks while still communicating. That is particularly important when we are coordinating effort in some collective manual task, like shifting a heavy or awkward object through a cluttered environment. It enables us to communicate over longer distances and in the dark. It is immediately attention-grabbing. It is less physically demanding, and it allows us to fully visually attend to and act on our environment while listening and talking (Irvine 2016). It is thus no surprise that humans everywhere primarily use speech rather than gesture to linguistically communicate unless forced to do otherwise. (In hunting, for example, silence is often important.) However, it is not enough to merely list those advantages. We need to detail a context in which those advantages were salient enough to drive a transition. We have already hinted at our pick for that context: the control of fire.

The control of fire provided a source of light after the sun went down. Light is the ordinary means by which animals set their daily routines. In great apes, the sleep-wakefulness cycle is controlled by the secretion of melatonin by the pituitary gland, which induces sleep. The visual perception of light inhibits the production of melatonin in the pituitary gland, and hence promotes wakefulness (Brainard, Hanifin, et al. 2001). Based on humans' current sleep needs, the control of fire would have on average extended the effective day length for hominins by as much as four hours in the tropics and even longer in winter in more northerly latitudes (Dunbar and Gowlett 2014). But firelight is both dim and spatially confined: a patch of relative brightness in a dark world. That limits the productive uses of time. After food was cooked, and a few tools were made or repaired, there would not have been much left to do other than socialize. Fire reduces the opportunity costs of social intercourse, while making it less optional. Without the extremely expensive option of everyone building their own fire, enjoying light and warmth enforces proximity. Cooking fires can be a small patch of hot embers, but light and warmth require more substantial fires.

Almost certainly, our ancestors came to control fire only gradually (Gowlett 2016). Fire would have been a regular and naturally occurring part of some hominin habitats. Fongoli chimpanzees inhabit a more open and arid environment than is typical for their species. These chimps are intrigued by fire, show an understanding of its dynamics on the landscape, and target burned areas for foraging after fire passes (Pruetz and Herzog 2017). Thus, early hominins likely acted similarly. This suggests that the first stage of hominin fire use was increased opportunistic use of wildfire. They might have actively followed wildfires, exploiting the presence of naturally cooked plant and animal foods (Parker, Keefe, et al. 2016). They may have spread existing fire, as some raptors do (Bonta, Gosford, et al. 2017). Animals fleeing from fire might have been hunted. Food might have been dropped into small, lingering pockets of fire or hot ash to cook or further cook them. Natural cooking very likely scaffolded more intentional cooking.

Increased opportunistic use of fire would have led to an enhanced understanding of fire. Harvesting wildfire and keeping it going would have been a natural next step. Fire maintenance is no trivial feat, however, and it is likely that it took hominins a long time to master this skill. Indeed, as Terrence Twomey (2013; 2014) has persuasively argued, fire maintenance imposes a range of cooperative, cognitive, and motivational challenges. To keep a fire going, we must constantly provision it. A working knowledge of what can be used as fuel, and where it can be sourced, is thus required. Hominins would have had to detach from the present, putting on hold current desires, to tend to this need. Gathering fuel would have been difficult and risky during the night. So it would often be important to collect fuel well before it was needed. Moreover, fire must be protected from rain and wind. Fire would have been particularly vulnerable when being transported from one location to another. The demands of fire on hominins would have been even more intense prior to the discovery of ignition technology. For one could not simply make fire on an as-needed basis, or recover it when it had accidentally gone out. However, as Haim Offek has noted, fire also creates a natural opportunity for the division of labor. The less mobile or skilled can still locate firewood, bring it to camp, tend an existing fire (Ofek 2001). Once a fire is in place and established, prior to the control of ignition it provides a focal point in the landscape and a reason (and perhaps a beacon) to return before nightfall. If Wrangham and others are right to argue that cooking is an ancient, erectine innovation, then a partial control of fire was one of the factors, together with reproductive cooperation, that rewarded central-place foraging. Moreover, fire-keeping is necessarily fallible, so if hominins really did have stable access to fire during this period but without ignition capacity, that suggests a network of fire keepers willing to exchange fire with one another (Gowlett 2006; Gowlett and Wrangham 2013). If residential groups were mutually suspicious in the way chimp bands are, this form of cooperation would be difficult to establish. Importantly though, once established it is intrinsically stable. Reciprocation is direct; help is cheap to give but very valuable to receive; and the flow of benefit is symmetrical. More ambitious uses of fire, as a tool for landscape management, are more cognitively demanding still. See Garde (2009) for an ethnography of the varied, demanding, and potentially dangerous uses of fire by Arnhem Land Australians.

The prehistory of ignition technology is probably quite complex, with this technology being discovered at different times in different places. It is also possible that it was lost only to be rediscovered at a later date. For example, in his recent review of the current state of the cooking hypothesis, Richard Wrangham acknowledges that there are surprisingly recent Neanderthal sites (within the last 200k years) with no evidence of fire (Wrangham 2017). As always, we must be cautious: sites are never entirely preserved. But especially when residential groups were thinly scattered without regular connection, technological loss would be no surprise. Suffice it to say that our ancestors likely controlled fire for an extended period of time before ignition technology was widely possessed, though perhaps imperfectly and with losses. Once the advance to ignition technology was made, fire use would presumably have begun to resemble its role in modern forager society. In the next section, we shall place the control of fire in its temporal context.

In our view, the control of fire was a major—if not *the* major—driver of the gesture-speech transition. We are not the first ones to link the control of fire with language evolution. Dunbar and Gowlett (2014) argue that human language was created by the fireside, using the increased budget of social time the control of fire made possible. Dunbar (2017) sees language as the final development in hominins' strategy for meeting the demands of social life. It allowed us to bond over the telling of stories and jokes. According to him, language evolved out of nighttime singing "by the very short additional step of mapping meaning onto sound" (2017, p. 211). Short step indeed! We do not agree. As we have argued through earlier chapters, quite sophisticated communicative capacities were needed to build erectine and heidelbergensian social worlds. That includes the firelight world. As with other aspects of the middle Pleistocene economy, the control of fire is demanding. It poses a range of coordination and cooperation problems. Solving these problems would not have required modern language. But a communication system more powerful than anything we see in other apes was probably necessary. Roles needed to be delegated (*who will watch the fire? who will retrieve the fuel? who will carry the embers with us?*), plans made (*shall we gather fuel now or later? what should we burn?*), real-time instructions given (*add this there! no, don't add that yet!*). These hominins probably needed cultural learning supported by communication to transmit the requisite natural history knowledge of fire and fuel. We doubt that our ancestors could have controlled fire, or managed other aspects of their economic life, without some significant progress toward language. We have argued that they had an upgraded communication system, and that it was primarily gestural. Some form of gestural protolanguage predated the firelight niche.

So, first, our information about erectine social and economic life suggests significant advances in their capacities to communicate. Second, if fire-using hominins only had great-ape-grade communicative abilities, the firelight niche would have been an unpromising environment for an initial expansion of communication. For the firelight niche lacks many of the ordinary props that facilitate interpretation and hence creation of signals. When individuals hunt together, the prey and its movements, together with those of the hunters, are all obvious topics, and so are likely to be the focus of communicative acts. Joint action is a potent mechanism of signal creation. But even just seeing what others are doing on their own, or simply where they are looking, can be highly useful. We read minds through behavior. Seeing that an individual is trying to extract a tuber from the ground with a digging stick facilitates insight into her current mental states. Knowledge of these mental states aids in communication, especially if she knows that you know what she is thinking.

In the firelight niche, by contrast, the effective communicative universe would have been initially much constrained. The visual world is smaller, less varied, less well illuminated, and fewer joint activities are in train. Objects that were present to hand (e.g., food items, tools) would have been available as obvious referents of communicative acts. Objects or persons located elsewhere and elsewhen (e.g., the stranger one saw today) would not. For those already fluent in displaced reference, objects, kinds, and places outside the shared visual field were feasible targets. But displaced reference was presumably not one of the first linguistic features to evolve, and it is difficult to imagine it evolving *de novo* around the fire.

However, while the fireside niche is an unpromising environment for *inventing* language, for those already competently using a multimodal but primarily gestural protolanguage, that niche would reward a shift toward greater reliance on talking. For example, if voice added redundancy to gesture in a mixed-modality system, the importance of voice could now increase, selecting for more readily distinguished vocal accompaniments to specific gestures. Likewise, many competences that evolved in the use of gesture would extend smoothly to a more vocal system. Turn-taking, for example, need not evolve anew. To the extent that enhanced working and semantic memory, enhanced theory of mind, and the capacity to understand displaced reference had already evolved to support gesture, these could also support a more vocal system. The control of fire then becomes an explanation for the gesture-speech transition rather than for the evolution of language itself. As Dunbar notes in passing, "gesture is difficult to make out across the half-light of the fireside, but spoken language carries far beyond from one hearth to the next" (2017, p. 11). Vocalization is very clearly superior in the dim light of the fireside. This is especially true if these fires were small and in the open air. Fire maintenance in the absence of ignition technology may well have been associated with smaller fires, as this would have conserved fuel, an important consideration if they have to be permanently alight. Dunbar is right that under such conditions "gesture is difficult to make out." Difficult enough that one might expect vocal elements to start to play an increasingly significant role in communication: perhaps to attract attention, perhaps to make displays easier to recognize and individuate. Those sitting right next to the fire would be able to converse using gesture, but it would be very hard for an individual at the periphery to be understood, especially if the visual world was not just smaller but more crowded, with more distant individuals partially occluded by closer ones. They might use vocalization more. It is also true, as Dunbar notes, that vocalization would allow communication between parties at different hearths. We suspect that the keeping of multiple fires might be a relatively recent habit, made possible by the control of ignition, given that without the control of ignition, fires must be maintained longer, increasing demands on fuel, but perhaps we are wrong. In any case, vocalization might still be used to communicate with an individual who has temporarily wandered away from the fire, to retrieve more fuel or to relieve themselves; here vocalization might serve as a kind of contact call. The more general advantages of speech over sign are valid in the fireside context: hominins could talk or listen while they were using their hands for other activities.

All this said, while gesture is more difficult to make out in firelight, it is not impossible. A gesturing agent can increase the amplitude of her signs; repeat them; slow them down. That is important, because it shows that the transition to a more vocal mode could be gradual and undisruptive. An established, mostly gestural protolanguage can still be used by firelight, just less efficiently. Firelight conditions are not optimal for gestural communication, but they certainly do not render it useless. We still gesture around the campfire at night. That allows gestural and vocal elements to work together in a transition. In order for arbitrary sounds to inherit meaning from co-produced gestures, those gestures must remain visible. Once an increased tendency to vocalize developed around the fire, perhaps at first paired with a sign that it helped disambiguate, increased vocalization could carry over into daytime hours, further reinforcing any gesture-vocal pairing and serving as a bridge to a more purely vocal system.

In sum, we think Dunbar and Gowlett are certainly right to draw attention to the firelight niche and its importance. It is difficult to overstate the consequences of the control of fire on hominin bodies, brains, and social lives. But unlike Dunbar and Gowlett, we call upon the control of fire to explain the gesture-speech transition, rather than the initial appearance of language.

6.6 Who, Where, When?

In our view, our ancestors began to predominantly talk rather than gesture around the fire. Which ancestors, and when? In this final section, we integrate our account of the gesture-speech transition with archaeological and genetic evidence for the evolution of speech. This serves as an important test for our account. We would be in trouble if the archaeological evidence of the control of fire, and the paleoanthropological dates for the capacities for speech, were wildly out of sync. Fortunately, these two fit together quite nicely.

That said, dating the control of fire is not easy. In open-air environments, it can be very hard to distinguish naturally occurring fire from anthropomorphic fire (fire created or maintained by hominins). To attempt to distinguish these, we must look to the broader archaeological context (Gowlett 2016). Artifacts or cut-marked bones in the same strata as burned material are suggestive of anthropomorphic fire. And if any of these items are burned, we can infer their co-presence with fire. However, these items might have been discarded by hominins at an earlier time and burned by wildfire only later. Hence, it is necessary to look at the distribution and ratio of burned to nonburned items on the landscape. If, for example, only a small portion of on-site flakes or bones show signs of burning, then that suggests a small, contained fire, which in turn suggests anthropomorphic fire. Micromorphological clues, such as the magnetic properties of burned sediments, can also provide information about the temperature and duration of fire. This is important, as anthropomorphic fires tend to burn hotter for longer in a single spot. The interpretation of such contextual clues is complex, and there is room for disagreement. Hence, many proposed anthropomorphic fire sites remain contentious. Other clues allow us to infer anthropomorphic fire with greater certainty. Flints whose functionality depends on hafting provide strong evidence of anthropomorphic fire, for hafting usually requires adhesives, which are made through the controlled application of heat. Direct evidence of such adhesives is stronger evidence still. However, these technologies doubtlessly took time to develop even after the control of fire, hence would not be present at the earliest sites containing domestic fires. So early claims of anthropomorphic fire are apt to remain the most contentious. Moreover, as emphasized by Gowlett (2016), direct traces of burning are few and fragile, creating a significant preservation bias. Stone tools persist for far longer than charcoal in most environments.

Anthropomorphic fire would have also occurred at lower frequencies than toolmaking, creating an additional sampling bias. With early control of fire, the absence of evidence is certainly not evidence of absence. Bearing this in mind, what does the fire record look like?

Briefly: there are a handful of open-air sites in Africa containing burned sediments, tools, and pieces of baked clay in the period spanning 1.5 to 1 mya. However, the status of these sites as loci of anthropomorphic fire, or even of opportunistic fire use by hominins, remains disputed. From about 1 to 0.5 mya, evidence for anthropomorphic fire becomes much stronger. This period contains the first unambiguous markers of hominin fire use. Fire sites are scattered throughout Africa, the Middle East, Europe, and Asia. Of particular significance is the open-air site Gesher Benot Ya'aqov in Israel, which contains many layers of charcoal, burnt wood, and burnt artifacts (Goren-Inbar, Alperson, et al. 2004; Alperson-Afil 2008; Alperson-Afil and Goren-Inbar 2010). The oldest layers containing such items go back to 780 kya. The presence of hearths is indicated by dense concentrations of burned artifacts localized in space. The period spanning 0.5 mya to 150 kya contains abundant evidence for the control of fire (relatively speaking) in all the same global regions. This is especially true after 400 kya. It is likely that this signals a change in fire-making/-keeping techniques. In particular, it would seem that larger hearths were being used. One site in Israel (Qesem Cave) shows evidence of a large, stone-lined hearth that was put to repeated use by hominins around 300 kya (Shahack-Gross, Berna, et al. 2014). (Even then, though, there are sites where we would expect to see signs of fire and do not: see for example Sandgathe, Dibble, et al. 2011.) Finally, as we enter the late Pleistocene, evidence for anthropomorphic fire suggests a role for fire on par with its role in modern forager societies.

Given our account of the gesture-speech transition, we particularly need to identify the period when hominins could reliably use fire, even if they still lacked the ability to ignite it. What does the record suggest here? A standard interpretation runs as follows. To the extent that the earliest fire sites (older than 1 mya) reflect hominin use of fire, this use was opportunistic. These hominins might have exploited naturally occurring fires, and even done so regularly, but they probably did not have the capacity to harvest, move, and keep naturally occurring fires going for indefinite stretches of time. The change we see in the record between 1 and 0.5 mya may well signal the arrival of control. The earliest hominins during this period were probably more novice than expert when it came to navigating the challenges of the control of fire. Thus, we would expect a gradual intensification of the role of fire in the lifeways of mid-Pleistocene *Homo*. The evidence of fire use at Gesher Benot Ya'aqov suggests hominins who were at the very least proficient keepers of fire. Finally, the trends we see in the fire record from 400 kya onward presumably reflect some combination of ignition technology and the more regular survival of these more recent traces (though it is possible that enhanced fire maintenance contributed as well).

Given this timeline, our account of the gesture-speech transition predicts that mid-Pleistocene *Homo* experienced selection for enhanced vocal communication. In particular, we would expect to see changes in *Homo heidelbergensis*, the last common ancestor of humans and Neanderthals, arriving on the scene around 800–700 kya. How does this prediction fare? In assessing it, we have two main sources of evidence to go on: fossil and genetic. There is mixed opinion regarding the potential for fossil evidence to tell us about the evolution of speech. Historically, proposed indicators have included cranial shape and dimensions, the size of the hypoglossal canal, the angle of the base of the skull, the morphology of the hyoid bone, and the width of the thoracic vertebral canal (Fitch 2000). Virtually all contemporary discussion revolves around the last two, the others having been dismissed as too unreliable.

Among other things, the hyoid bone supports the larynx. It is widely accepted that the Neanderthal hyoid is essentially humanlike. This suggests that the hyoid had already evolved into its modern form in heidelbergensians. Fossil remains from the Sima de los Huesos site in Spain are consistent with this (Martínez, Arsuaga, et al. 2008). Two heidelbergensian hyoid bones have been recovered at this site and both show an essentially modern morphology. Some doubt whether the hyoid bone provides much information about speech abilities; in particular, it has been criticized as an indicator of the location of the larynx in the vocal tract. Skeptics (most notably Fitch) point out that there is no change in hyoid morphology over the course of human development, despite its pronounced descent. However, it is uncontentious that the structure of all Neanderthal and heidelbergensian hyoids reveals the loss of the air sacs early hominins shared with *Pan*, a clear step in the direction of modern speech anatomy.

It is interesting to compare the modern-looking hyoid of *heidelbergensis* with that of erectines. Erectine hyoid morphology reflects an intriguing mixture of modern and archaic traits. As described by Capasso, Michetti, and D'Anastasio (2008), the erectine hyoid assumes the bar shape characteristic of later *Homo* as opposed to the bulla shape characteristic of *Pan* and *Gorilla*. The bulla shape, but not the bar shape, is assumed to accompany air sacs, suggesting that this ancestral trait had already been lost in erectines. On the other hand, in a number of respects the morphology of the erectine hyoid more closely resembles that of chimpanzees than of later *Homo*, leading these authors to conclude that *erectus* probably did not have modern speech capacities. In particular, the hyoid shows no evidence of attachment to an important muscle group (supra hyoid muscles) which in humans functions to modulate the high end (upper regions) of the vocal tract.

As for the other putatively reliable fossil indicator of speech capacity: the width of the thoracic vertebral canal is taken to indicate the degree of breath control. Abdominal and intercostal muscles are controlled by motor neurons that reside in this canal. Hence, a thicker canal presumably implies enhanced control of these muscles. Based on current evidence, the thoracic vertebral canal of *erectus* seems to fall within the modern human range (Villamil 2014). However, as has been pointed out—again by Fitch (2010)—that enhanced breath control might have evolved in the service of running or even swimming. In light of our discussion in section 6.4, to this list we should add laughter and singing.

An upgrade in speech capacities predicts corresponding changes in auditory apparatus; there is little point in producing a range of new vocalizations unless these sounds can be discriminated by hearers. In this area evidence is limited, but the evidence we do have paints a clear picture. Fossil reconstructions of the external and middle ear in *Homo heidelbergensis*, again from the Sima de los Huesos site in Spain, suggest essentially modern auditory capacities. In particular, these authors (Martínez, Rosa, et al. 2004; Martínez, Quam, and Rosa 2008) show that the reconstructed anatomy of the hearing canal implies increased sensitivity around 400 Hz compared to chimpanzees and other great apes. This inference is further substantiated by evidence from the Middle Paleolithic allowing us to compare Neanderthal and modern human ear ossicles (middle ear bones) (Stoessel, David, et al. 2016). This evidence shows that the morphological properties of the former fell within the modern human range. Thus, it would appear that auditory capacities capable of registering modern speech were in place before the split between Neanderthals and humans, and inherited by both lineages from an older ancestor.

What about the genetic evidence? DNA comparisons between Neanderthals and modern humans have revealed a striking degree of similarity. To date, a total of 96 amino acid substitutions have been identified in 87 proteins (Prüfer, Racimo, et al. 2014). Noncoding differences are more substantial, with current estimates of around 3,000. Noncoding DNA differences are not as well understood, but at least some likely play an important role in up-regulating or down-regulating gene expression. Some of the genes showing coding differences or differences in nearby noncoding DNA are expressed in the brain and so may well have important effects on language and speech. However, many others seem to be unrelated to language, affecting such things as skin, hair, bone, immunity, and reproduction. One important general point has been emphasized by Levinson and Dediu: given the size of the human genome, there are rather few differences between *sapiens*, Neanderthals, Denisovans, and by inference their common ancestor, *heidelbergensis* (Dediu and Levinson 2013; Levinson and Dediu 2018). We are genetically *very* similar to our Neanderthal and Denisovan siblings, and by inference to our *heidelbergensis* ancestor.

Much attention has been paid to the gene designated *FOXP2*. It was once thought that *FOXP2* was critical to human syntactic capacities, and hence it was dubbed the "language gene." The more recent understanding is that it makes possible precise vocal control. The coding region of *FOXP2* is identical in humans and Neanderthals. However, there is a difference in one of the upstream binding sites for a transcription factor (*POU3F2*). It might be tempting to think that this difference is relevant to speech. But as many as 10% of modern Africans show the archaic allele at this position, and so, clearly, this mutation is not required for modern speech (Maricic, Günther, et al. 2013). Other regulatory DNA differences associated with genes implicated in language and speech have been identified (for example, differences in regulatory DNA controlling *ROBO1*, *ROBO2*, and *CNTNAP2*)[9] but the phenotypic effects of these changes are yet to be firmly identified. Moreover, there remains much to learn about extant genetic variation and developmental plasticity in modern humans that might bear on these issues (Dediu and Levinson 2013).

In sum, then, while the genomes of modern humans and Neanderthals are strikingly similar, a number of changes have occurred in our line since our split with the Neanderthals some 500–700 kya. Of these changes, a small number may be implicated in the development of human speech capacities. This idea is further reinforced by some recent work showing epigenetic differences between modern human and Neanderthal genomes (Gokhman, Nissim-Rafinia, et al. 2019). Specifically, it has been shown that in our lineage new patterns of methylation have evolved on a suite of genes involved in facial and vocal tract morphology (*SOX9, ACAN, COL2A1, NFIX*, and *XYLT1*). But on the whole, there is little to suggest a major architectural difference between human and Neanderthal vocal tracts or neuroanatomical structures relevant to speech. Rather, these differences appear more like "finishing touches" to an already sophisticated system supporting speech production and perception capacities, a conclusion further supported by the fossil evidence mentioned above. Moreover, differences in *sapiens* speech compared to that of Neanderthals need not mean that Neanderthal capacities were limited or impaired compared to those of *sapiens*. All things considered, the evidence suggests that the capacities for speech were present in the mid-Pleistocene ancestor of humans and Neanderthals. The capacity for speech was probably available to these hominins, the heidelbergensians, as they settled into the fireside niche.

Thus, both the archaeological and genetic evidence fits with our account of the gesture-speech transition, though the dates are far from iron-clad, and the evidence remains fragmentary and open to interpretation. Are we to conclude from this that Neanderthals, and before them heidelbergensians, possessed an essentially modern form of spoken language? Not quite. Rather, it shows that they very likely possessed the requisite vocal and auditory machinery to support such a communication system. They may well have had the requisite cognitive capacities too. They were probably language-ready and speech-ready. But in our view, the full suite of features that mark out modern language as a distinct type of communication system relative to earlier protolanguages probably did not emerge until sometime in the last 200k years. In our view, those features were both a symptom and cause of the increased social complexity of modern sapient lifeways. Specifically, we claim that language as we now know it emerged contemporaneously with the second of our cooperation revolutions—the reciprocation revolution. Prior to this revolution, we doubt there would have been a genuine need for the full range of communicative resources made possible by modern language. Modern language depends on features of the social environment, not just on individual cognitive or structural machinery. We unpack this argument in the next chapter. So the next chapter will see the final iteration of the inferential strategy that threads through this book: a case that our ancestors had the cognitive and social resources that would support an enhanced form of communication, paired with a case for the view that their social and economic lives would have been much enhanced by that form of communication. By 200 kya, and likely much earlier, ancient humans had the intrinsic cognitive resources required for full language. But they may well not yet have had a social life that would support the maintenance and transmission of full language. Furthermore, we do not think their lives at 200 kya were economically, socially, or technically complex enough to require these rich communicative tools. As chapter 7 will show, both these conditions begin to change from very approximately that date.

7 From Protolanguage to Language

7.1 The Changing Face of Cooperation

We have argued that the evolution of language was a long, gradual process. That process began with an elaboration of great ape communication, particularly great ape gestural communication. By the time of *erectus*, cooperation in our line had expanded very considerably. *Erectus* likely practiced cooperative breeding, foraging, hunting, toolmaking, and information-sharing. These developments drove advances in hominin communication systems which fed back to further transform cooperation, creating a positive coevolutionary feedback loop.

In our view, the ecology of early hominin cooperation was predominantly immediate-return mutualism. In this form of cooperation, agents act together to deliver some desirable outcome, a cooperation profit. That profit is (or can be) then divided on the spot. Collective defense against a predator is an immediate-return mutualism of this kind; when successful, all the agents in the coalition have come through a risk unscathed. So too is collectively driving a predator from its prey, which is then available to be shared. Michael Tomasello is well known for arguing that this was the foundational form of human cooperation, and we agree (Tomasello, Melis, et al. 2012; Tomasello 2014). Immediate-return mutualism is cognitively and motivationally simpler than other forms of cooperation, whose benefits depend on planning for the future, and/or trust that others will repay in the future favors done for them by donors now. Immediate-return mutualism does not make cheating impossible, but it does make it obvious if and when it happens. If an individual hangs back while others chase a predator away from a carcass, his defection is plain for all to see. Being part of a cooperative enterprise does not require passing up a benefit now in the expectation of a greater reward in the future, as, for example, sharing your food now in the expectation of future help does. Likewise, punishing those who cheat requires only simple cognitive and motivational mechanisms. In favorable cases, punishment might involve no more than denying the freerider access to the fruits of cooperation. The individual who hangs back might be blocked from the carcass until all its good parts have been stripped, for example. If so, then there is little or no point in freeriding. Even if more active measures are required against, say, a cheat who tries to hijack the whole carcass, everyone else present stands to lose, and so all will have some immediate and affectively powerful motivation to intervene (these ideas are developed in more detail in Sterelny 2021). However, while we think earlier forms of cooperation were predominantly versions of immediate-return mutualism, there were probably exceptions. We think the most important was cooperative breeding, involving cooperation amongst close relatives and/or direct reciprocation: caring for another's baby in the expectation of a return favor in short order.

These forms of immediate-return mutualism facilitated the evolution of more motivationally and cognitively demanding forms of cooperation. One piece of evidence for this is the incremental expansion of the hominins out of "Savannastan"[1] over the last million years or so, as they became increasingly capable of flourishing in a broader range of habitats, as documented in, for example, Gamble (2013), Finlayson (2014), and Roberts and Stewart (2018). As cooperation became increasingly central to hominin lifeways, there would have been selection for traits enhancing judicious cooperation. Most obviously, these would have included greater prosocial tendencies for active cooperation, including active cooperation against bullies and freeloaders, but also greater social tolerance. Over the course of the middle and late Pleistocene, cooperation in our line continued to expand. As we see it, cooperation changed in three ways. First, it expanded in scope. More of the routine activities of daily life came to involve cooperative interactions with others. We see this with fire: as we explained in the last chapter, this required hominins to solve a range of cooperation and coordination problems. A second is a change in the social and physical scale of cooperation. The social world of a great ape is more or less confined to its residential group. Very likely, that was the ancestral hominin pattern too, but it is very much not the pattern of ethnographically known foragers, let alone other human cultures (Seabright 2010). Forager bands are open, in contact with other bands, with many recognized relations between them, and typically with fairly free movement between bands belonging to the same tribe or clan (see for example Lee 1979; Marlowe 2010). While this change is very difficult to date and explain,[2] at some period in hominin evolutionary history hominin residential groups gradually become more open, with individual and collective cooperation across bands. By the Holocene, there is evidence of quite large-scale constructions of drive lines, fish weirs, and fish traps: constructions well beyond the capacity of a single residential group (see Frison 2004 for impressive North American examples). This increase in social scale—the fact that bands were not socially or informationally isolated islands—made cumulative social learning much more stable (as in small groups innovations can be easily lost through an unlucky death). And probably faster, as good ideas can be borrowed from the neighbors. As we have often remarked, language is a challenging social learning target, and the richer and more complex the language, the greater the challenge. This increase in social scale may have been essential for the emergence of full language (Planer 2020a).

A third development, and this will be our main focus, is a change in the economic organization of cooperation. Our ancestors increasingly engaged in forms of reciprocal or delayed-return cooperation. These grade shifts in cooperation changed the socioeconomic basis of hominin life in profound ways. They all offered net benefits. But they also introduced novel social conflicts. The more complex various cooperative practices are, the more there is to go wrong.

These changes in the form and scope of cooperation are relevant to the evolution of language, for in our view, the need to control these novel conflict risks and lower their costs was a critical driver of the evolution of full language. In developing this idea, we first consider a range of complex forms of cooperation. That identifies a set of potential problems, conflict flashpoints. We then identify a set of social tools that mitigated these problems. No solution is perfect, but these social innovations reduced these conflict costs to tolerable levels.

However, these solutions all involve sophisticated forms of language. It is possible that Neanderthals or even some earlier species of *Homo* also practiced these forms of cooperation, or some of them. They are not a package deal. If so, then our account would predict that these other hominins also possessed some or all of the ingredients of the package of design features we take to mark out full language. However, based on current evidence, but with due caution, we see no reason to think that any other *Homo* species cooperated in all of the ways that late Pleistocene humans did. Hence, toward the end of the chapter we also take up the question of what triggered these uniquely *sapiens* forms of cooperation. Why only us? If only late Pleistocene anatomically modern humans (AMHs) had full language, we need at least a tentative explanation of that fact. We propose it was because only those humans cooperated so extensively. Without committing to a specific hypothesis, we sketch some possibilities, some potential explanations as to why the most extensive forms of cooperation were confined to late Pleistocene AMHs.

7.2 The Social Costs of New Ways of Cooperating

The greatest accomplishments of humans, whether in the ecological, technological, or social domain, depend on a foundation of cooperation. What is especially remarkable about humans is the extent to which we cooperate with nonrelatives, and in the limit case, with total strangers (Seabright 2010). But there are some real social costs to being super-cooperators: conflict costs and risk costs. In this section, we highlight some of those costs, and how they emerge from forms of cooperation that potentially have a net benefit despite these costs.

7.2.1 The Division of Labor

A division of labor is a universal feature of forager life. The most common form of division of labor is along sex lines (Kelly 2013). Adult men and women target different resources over the course of the day. Men typically hunt and sometimes fish, while women gather a variety of plant-based resources and/or small game such as insects and lizards. When hunting is successful, meat is shared liberally across the band. Women primarily provide support to their families, including their male partners (though there are cultures in which all food is shared). Hunting would not be feasible were it not for the support of women in these societies, as hunting commonly fails. In some forager societies, there is also some age-based division of labor, with older males (especially) devoting more of their energies to ritual, storytelling, and similar social goods (Hart and Pilling 1960; Lewis 2015).

This division of labor reduces risk. For to the extent that the band no longer forages as a single unit or with a single target, its members are no longer bound to succeed or fail together (Cashdan 1980). The band is made less vulnerable in the face of stochastic fluctuation of naturally occurring resources. Such a division of labor facilitates specialization. Men and women (for example) acquire the physical skills needed to make, use, and repair different sets of tools. For some tools, acquiring these abilities takes considerable time and effort. Making and using a bow is far from easy. Specialization also makes it worthwhile to invest time and effort in making specialist equipment, like fishing nets or fish traps. For foragers, nets and traps are typically big-ticket items (Kelly 1984; Satterthwait 1986; Satterthwait 1987). They are only worth the costs if these items are used regularly. There is also cognitive specialization. Skill at hunting requires a very different knowledge base than skill at gathering. Hunted and gathered resources are distributed differently, and follow different time schedules. There are different cues one must learn to read. The development of expertise in these domains depends to a large extent on cultural learning, but also on prolonged trial-and-error learning by the individual. We learn by doing in a physical and cognitive environment constructed for us by experts (Sterelny 2012a). Division of labor makes it possible for an individual to forage successfully without mastering the entire body of information possessed by his or her community as a whole. At the same time, the group as a whole comes to possess know-how and know-that which no single individual—even one supported by extensive cultural learning—could reasonably be expected to gain in a single lifetime.

The division of labor with its associated specialization thus tends to increase the efficiency of foraging in important ways. It tends to increase the resource breadth of a group. This tends to reduce risk, as it is unlikely that all sources of supply will fail at once. If the residential group forages by fractionating into small parties, the habitat around base camp is searched more efficiently (more on this later). A larger fraction of the potential resource pie is taken. But with this efficiency comes new potentials for social friction. If risk is managed by sharing, with those who succeed today sharing with those who fail, then those who share must trust that they will receive fair-value return favors in the future, and there can obviously be different views about what counts as fair value. That is especially true when we consider individuals' natural biases in valuing their own contributions. These tensions are likely to be magnified if there are differences in skill and commitment that result in some needing to share others' successes much more often than they have their own successes to share; and such differences are surely likely. They are also likely to be magnified by problems of commensurability. The efficiency of a foraging division of labor in part depends on broadening the resource base, but that brings with it the problem of exchange rates. If one individual regularly contributes fish, what counts as a fair return of other resources (Sterelny 2014)? One party might feel it is appropriate to pay back a waterfowl with a basket of shellfish, whereas the other might feel shortchanged. More generally, as the number of goods in circulation in a group increases, there is increased opportunity for conflict over what constitutes acceptable exchange. That is true not just for the exchange of food, but also for tools or hybrid forms of exchange (e.g., the exchange of a javelin for a leg of reindeer, or food for services). Conflict can even extend to the exchange of information. Such conflicts threaten continued cooperation among group members, or worse.

Harvesting a greater range of resources can strain social life in another way. For different resources become available at different times and places, and with very differing degrees of certainty. From the perspective of those targeting one resource, relocating camp further inland may be optimal. Low-value resources are not worth carrying long distances, so those who specialize in these (often women) will want to move camp more often than those who target high-value resources. With more abundant but lower-value resources, it is efficient to move the group to the resource; with less abundant, less predictably found, higher-value resources, it is more efficient to move the resource to the group (Binford 1980). Moreover, transportation costs can also be unevenly distributed across the group. If it is women who are responsible for carrying infants and small children (as is usual), then those costs may be significantly higher for women than for men. If there is cumbersome kit to transport, then this too can create divergent interests. In addition to these material costs, there are motivational stresses too. Thus, as group members come to specialize in different resources, there is a tendency for them to more widely disperse over time and space, and that can include specialist parties traveling for days to a location where there is a particular resource, but which is otherwise unattractive; for example, nesting colonies on offshore islands, which often have very limited fresh water. The loss of daily contact can lead to an erosion of trust and cooperative motivation. In particular, the separation of male and female partners is likely to intensify sexual conflict, as males have a much-reduced ability to monitor and physically guard their mates from other males (Sterelny 2019a). So decisions about when and where to move become more fraught, and paternity uncertainty is increased.

7.2.2 Reciprocity

When cooperation takes place in the here and now, the motivational and cognitive demands of cooperation are kept to a minimum. Conversely, when cooperation becomes spread out over time, those demands increase. That is true even in the simplest case of direct reciprocation, where one individual helps another at one moment in time, expecting a return favor later. In helping, the individual detaches from the present and behaves with an eye toward the future. As we all know, it can be challenging to forgo a current benefit for one in the future, even if we expect that future benefit to be greater than the one we could take now. Likewise, the individual helped must be motivated to return the favor, again despite perhaps strong temptations to let the future take care of itself and keep what he or she now has. In addition, individuals must remember favors given and favors returned, and that can be difficult with longer time horizons. These cognitive and motivational challenges become significantly more difficult when cooperation depends on indirect reciprocation: when the help given is apt to be compensated by some third party, and perhaps in a different form. For then ensuring fairness—a reasonable balance of favors given and received—depends on tracking interactions over the whole group, not just between a focal agent and those with whom that agent directly interacted. To know that Thag deserves help when he returns empty-handed, Thug needs to know that Thag helped out Thagma last week, even though he did not have much himself. For this reason, it is much more difficult to detect freeriders—those who regularly contribute less than they receive—in systems of indirect reciprocation. It is also much more difficult to assign responsibility if one individual is consistently on the wrong side of the balance sheet: if, say, an individual who regularly contributes is ungenerously compensated. For who, exactly, should have given more? Likewise, it is less clear who has the responsibility to bear the risks of punishment, if freeriding is detected. Yet reducing risk by sharing certainly works best with indirect reciprocation:

whoever succeeds today shares with those who failed. These features make reciprocal cooperation much more fragile than immediate-return mutualism. Reciprocity is fraught with potential conflicts and stressors.

The life of humans in the late Pleistocene came to feature many forms of reciprocity. Rob Boyd points out that ethnographically known foragers exchange many services, not just material goods, and this is likely to have had ancient roots, not least in the exchange of services with cooperative breeding (Boyd 2016). However, while some reciprocal cooperation very likely has these deep roots in hominin history, we suggest that the role of reciprocation became much greater in hominin life with the evolution of projectile weapons technology. *Erectus* very likely used strong, stabbing spears, or short-range spears, to kill medium and large game. Evidence of *erectus* shoulder morphology suggests that erectines could execute fluid, powerful throwing motions (Roach and Richmond 2015). Thus it is quite possible that erectines made use of some kind of projectile technology, whether short-range throwing spears or lithic missiles.[3] At Schöningen in Germany, throwing spears have been found that date to around 400 kya. The aerodynamic properties of these spears suggest that they were short-range weapons (though perhaps still effective at 20 meters), and they were probably used by European *heidelbergensis*. Currently, the earliest evidence we have of composite projectile weaponry is stone-tipped wooden spears from Ethiopia. These spears date to a minimum of 279 kya (Sahle, Hutchings, et al. 2013). The microfracture pattern of the tips suggests higher impact velocities than would be achievable via thrusting. This is an intriguing find, as there is no other clear evidence of composite projectile technology for at least another 150k years (Churchill and Rhodes 2009). Stone-tipped throwing spears are followed by the invention of the atlatl (a spear-throwing lever) and the bow and arrow. These more sophisticated projectile technologies more clearly appear in the African archaeological record after 100 kya (Shea 2006; Shea and Sisk 2010; though O'Driscoll and Thompson 2018 suggest a slightly earlier date).

Once high-velocity projectile weapons are available, we expect to see the size of hunting parties shrink, as a single hunter (or a very small party) has enough firepower to kill, and a small party can stalk or lie concealed much more efficiently than a large one.[4] Forager hunters do indeed typically hunt alone or in small groups. This advantage of stealth helps in hunting large game, but is especially helpful in hunting small game, as these animals tend to be more timid. As a consequence, if men hunt alone or in groups of two or three, a typical band can support multiple hunting parties. As we said in section 7.2.1, this increases the overall efficiency of search and reduces risk. For it greatly increases the chance that at least one of the parties will encounter prey. But in these circumstances hunting becomes part of an economy of reciprocation, taking the group away from immediate-return mutualism. Before projectile weapons, hunting parties had to be bigger. Large animals pose serious danger to hunters armed only with stabbing spears, and there is strength in numbers. Multiple, coordinate strikes were very likely needed to take down such animals at acceptable levels of risk.

So while hunting with projectile weapons increased the efficiency of hunting, it opened the door to new conflict points. Once hunters form small, dispersed groups, they will have varying success on any given day and over time. Reciprocation, if it is reliable, is certainly beneficial, as no one can be certain of success: a regular report from ethnography is that hunts often fail. It is beneficial to share meat with others if doing so will induce them to share with you (and your family) on days when you are unsuccessful (or less successful). This is not to say, however, that there would not be patterns in who contributes what. For while success at hunting does depend on luck, it also depends on knowledge and skill. Not everyone is an equally talented hunter, and those who are more talented will contribute more meat to the group. They are more valuable group assets, and they may well know it, and attempt to leverage their success to gain other assets; for example, sexual access. Indeed, there is a lively debate in the ethnographic literature on the importance of such leveraging in explaining male hunting (briefly reviewed in Jaeggi and Gurven 2017). This can breed social ill-will. Even if they do not secure a threatening amount of sexual access, those who are more successful may begin to feel superior and, if left unchecked, may attempt to exert control over other group members (to "big note" themselves). This tendency must be counterbalanced in some way if group cohesion is to be preserved.

In short: we suggest that the differential success of small hunting parties played a critical role in making reciprocity a central feature of human lifeways. But while this is a very profitable form of cooperation, it is much more cognitively and motivationally complex than immediate-return mutualism. If and as these problems in managing reciprocation stresses in one domain were successfully met, that probably facilitated reciprocity in other domains.

7.2.3 Cooperation across Bands

Forager society is complex, with bands nested in larger communities. Minimally, a number of bands combine to form a cohesive metapopulation or metaband—typically, an ethnolinguistic unit. A band is often associated with a particular territory. However, band membership is fluid (Hill, Walker, et al. 2011). For example, a grandmother might go to live in her daughter's band following the birth of a grandchild. Alternatively, an adult male might relocate with his entire nuclear family to another band for an indefinite period. This is one response to conflict and disagreement (Marlowe 2010). Moreover, social life is outward-looking, with individuals in one band having kin, friends, and other social allies in other bands (Wiessner 2014). Forager society strongly contrasts with *Pan* society in this regard. A lone chimp is very likely to be killed if he encounters a group of males from another residential group. The same is true of a female chimp and (if she has any) her offspring (Stanford 2018). Intergroup relations are less hostile in bonobos than in chimps. Female bonobos often embrace extragroup females and engage in genito-genital rubbing with them. Females also copulate with extragroup males. Male-male relations remain tense, but males generally do not attack one another (Furuichi 2011). Notwithstanding these differences, bonobo society remains closed in an important sense. Neither males nor females are permitted to switch residential groups at will. As with chimps, males are constrained to live out their lives in their birth group, while female migration is generally limited to an initial period of dispersion.

Humans cooperate across bands in a variety of ways. Individuals travel between bands within their larger metaband to exchange food, tools, and raw materials with one another. Information about environmental resources is exchanged (Whallon 2011). People themselves are exchanged, or just shift. Women are frequently exchanged in marriage. If one band is currently experiencing a resource shortage, they are often allowed onto the territory of another band (Kelly 2013). During periods of seasonal plenty, several bands might gather together for days or weeks. In Australian Aboriginal life, these were the occasions of intense ritual activity (see Spencer 1928 for vivid descriptions and amazing photos that show the extraordinarily elaborate and intricate costumes of the participants, and the great length of some of these ceremonies). This is a critical opportunity to renew existing cooperative relations and forge new ones. All of this works to provide further social insurance to foragers by distributing risk over multiple bands.

As we have already seen, it also has important consequences for cumulative cultural evolution, as innovations more easily stabilize in the presence of robust information flow between groups (Powell, Shennan, et al. 2009).

There is a serious problem in explaining how social organization shifted from the closed social worlds of the great apes to these more open forager worlds. But even once the basic framework of an open social life was in place, metaband organization is not cost-free. It introduces a range of conflicts and social tensions. The more people are present, the more interactions there are, the greater the chances of social friction: individuals take a dislike to one another, an old grudge resurfaces, new ones emerge in the face of new sexual temptations, there are squabbles about food (none of these uncommon, according to Boehm 1999 and Wiessner 2014). In these cultures without formal leadership, there are no institutional mechanisms to prevent hot blood escalating to violence. Murder rates in forager societies are high.[5] If violence does erupt, that can generate further stress within and across bands, as the contending parties seek social support. For within a band, individuals have overlapping but by no means identical commitments to kin and allies in other bands, so conflict without can cause divided loyalties within. If there are resource shortages, again loyalties can be divided. One band member may wish to provide resources to another in which she has family. Her bandmates might disagree. That is especially likely during periods of general environmental stress such as drought. These divergent interests pose an obvious threat to relations within the band and hence to band cohesion.

In addition to the specific benefits of information and demographic exchange, and in a few cases large-scale collective action projects, community-scale organization can increase the scope of reciprocation and the division of labor. As the number of individuals increases, there are opportunities for further division of labor and specialization. For as pointed out by Ofek (2001), specialization depends on market size. So long as exchange remains within the band, there is limited scope to support specialist toolmakers. An individual specializing in the production of arrows will quickly exhaust the needs of his group. That becomes less true once forager society becomes open and there are opportunities for trade (Marwick 2003); such trade can even involve cultural as well as material goods (Lewis 2015). As before, while cooperation becomes more potentially valuable, the threat of conflict also increases. That is especially true of reciprocal relations with individuals in other bands, as those relations will in general have a longer time horizon than reciprocal relations within the band. Thus, the social challenges posed by reciprocity are exacerbated.

7.3 Social Stress Relief

We have argued that as these new forms of communication emerged and expanded, human social life grew increasingly complex and conflict-prone. In response, humans developed a range of strategies for avoiding and controlling conflict. In this section, we consider three such coping mechanisms. Each poses strong demands on linguistic communication. We propose that the development of these mechanisms was a critical part of how and why humans evolved full language.

7.3.1 Norms

Norms are *charged* expectations. As Peter Railton expresses it, when we see others violating one of these expectations, we have some impulse to protest or disapprove, even if our own interests are not affected, and even though we sometimes do not act on those impulses. When we notice that we ourselves have violated one, we feel some emotional discomfort, even if no one else has noticed; even if no one else is affected (Railton 2006). These negative feelings tend to motivate us to make amends for the violation and to avoid violating the norm again. These tendencies (or at least clear precursors to them) are evident from quite an early age, suggesting some form of biological preparedness in the normative domain.[6] Some norms are so deeply held that we are moved to intense anger when others violate them, leading us to support punishing the offender. This can be true even of apparently unimportant norms. When Sterelny was young (in the Pleistocene), many adult males were outraged by the sight of young men with long hair. Even norms like these can be very deeply felt. The term "norm psychology" is sometimes used to describe this constellation of psychological facts about humans.

Norm guidance is often unconscious and automatic. Some of the norms that guide us are implicit, never articulated but nonetheless powerful. Conversational norms of loudness, distance, and turn-taking only become obvious when you visit cultures in which they differ.

Others are or were explicit, and many of these norms would be very unlikely to arise and/or stabilize without the ability to make them explicit. For example some foragers have remarkably complex norms for dividing large game: two examples are Gould on kangaroos in the West Desert and Alvard and Nolin on whale carcasses (Gould 1969; Alvard and Nolin 2002). Other examples are norms about social interactions: in many Australian Aboriginal societies there is a norm specifying that one is not to be alone with one's mother-in-law (or even someone who might legitimately have been your mother-in-law); likewise, in many of those cultures there are norms proscribing the use of the name of the recently dead. Without language, a learner would struggle to identify these rules. While these norms are usually respected, and so result in regularities of action, these regularities are only visible when agents are identified using a complex set of kin terms.

We think the development of more elaborate norms was an important mechanism through which humans met the new social challenges they faced (as argued in Zawidzki 2013). As Robert Boyd (2016) points out, they help to make ambiguous social situations—situations in which there might be a difference in views about whether agents have cooperated—less ambiguous, by making community expectations explicit. In this way, they help bring agents' expectations into mutual conformity. If it is a norm that a duck is to be repaid with two pigeons, then an agent who has supplied a duck will be expecting two pigeons, and the agent who has received the duck will be expecting the other agent to expect this. In making mutual expectations more explicit, they also simplify the monitoring problem. In the absence of a norm specifying how a duck is to be repaid, a good-faith attempt to reciprocate might be mistaken for an attempt to cheat. An attempt to pay back a duck with two catfish might be perceived as unfair by one party despite the fact that the other party fully intends his offer to be fair and cooperative. Failing to distinguish misfires from genuine cases of defection is itself a conflict flashpoint. For agents who are genuinely being cooperative deeply resent punishment being directed at them: behavioral economics experiments show that this can trigger retaliatory punishment cascades. Moreover, norms make it easier for an individual to enlist third-party help. For the aggrieved has an agent-neutral standard to appeal to: norms of division, or norms of marriageability that define an agent's possible partners, apply to classes of agents. For example, Australian Aboriginal communities are divided into moieties, and no agent in a moiety can marry within that moiety. That marriage prohibition is agent-neutral. As a strong norm psychology came online, such appeals would have been increasingly effective. Finally, norms will tend to promote group agreement about punishment.

That is especially true if norms specify the punishment for norm violation, or how a decision about punishment is to be reached and who makes the decision. Norms of this degree of complexity and completeness help avoid higher-order conflict about the right response to norm violation, but their complexity and completeness require the full resources of language.

This dependence on language comes out particularly clearly in Robert Boyd's account of norm enforcement amongst the Turkana, an African pastoralist culture that engages in a good deal of cattle raiding. This has the potential to be seriously dangerous, and hence sometimes results in commitment failures in the raiding party. Such failures are punished, with varying severity; Boyd says of this, "Community consensus determines how someone is sanctioned. Local people, especially age-mates, discuss the violator's behavior. Once a consensus emerges, members of the violator's age-group are responsible for administering punishment, even if they did not participate in the raid and did not experience the consequences of the violation" (Boyd 2016, p. 96). As we can see, the social process through which sanctions are inflicted is very culturally and linguistically rich.

Some norms directly encourage cooperation by creating mutual obligations; a classic example is gift exchange, as described by Wiessner (2002); but norms need not directly mandate cooperation in order to help keep the peace. Norms can encourage ways of interacting socially which increase cohesion and thereby indirectly reduce conflict. A nice example is provided by Wiessner (2014). She reports a dramatic shift in conversational topic among the !Kung San between daytime and nighttime hours. Specifically, 34% of daytime conversations consisted of agents complaining about one another (often about stinginess). In contrast, only 7% of nighttime conversations were complaints, while 81% of the time agents told entertaining stories. This tendency to refrain from criticism during nighttime hours and instead enjoy storytelling with one another helps deflate social tensions that have built up over the course of the day.

As human social life came to be saturated in norms, particularly explicit and complex norms, we would expect growing linguistic sophistication. Norms are typically stated in an abstract way; they govern the behavior of agents, considered impersonally, or of particular *kinds* of agents (e.g., a son's obligation to his wife's parents). They often involve categories that can probably only be learned with something like language: one's cross cousins, or one's totemic animal. Norms have deontic force: they tell us what we *can* or *must* do. This takes communication away from simple indicative and imperative communication. To express a norm, we must be able to express mood: ought, must, permissible, forbidden. We must be able to express conditionals, often complex ones: if X does p, unless q, then Z. We must be able to express counterfactuals in order to assign causal responsibility: if Thag had kept the fire going, Thagma would not have got lost. It is very hard to imagine how complex norms could be expressed and communicated without quite complex syntax: you cannot be alone with your mother-in-law unless a third party is present. Likewise, specifying who should benefit when a rich prize is divided in an impersonal way seems to require structured referring expressions: the hunter's maternal uncles are entitled to share the shoulder of a kangaroo. In addition, as social life becomes more dispersed in time and space, from the perspective of any given agent more and more happens offstage, away from direct observation. Increasingly, members of a band will have to rely on reports of others' behavior in the elsewhere and elsewhen. In other words, gossip becomes central; more on this in section 7.3.3. Finally, while norms increase the transparency of social interactions, ambiguities and marginal cases will always remain, as will issues about the severity of the norm violation (if any) and the availability of reasonable excuse. There is bound to be disagreement here, with different proposals about whether the norm was violated, and what the appropriate response should be. Mercier and Sperber (2017) have proposed that public reasoning—defending one's own views and criticizing those of others—became important in contexts like these. We find their proposal persuasive, but it too presupposes rich linguistic capacities: metarepresentational tools which enable us to represent and assess what others have said (and what we have said in the past).

7.3.2 Kinship

Human kinship is not biological kinship. This is most evident in the recognition of affinal kin as kin—a husband or wife and his or her relatives. But even consanguineal kin relations do not neatly map to biological kin relations. At the same time, they do not float freely from biology. Consanguineal kin relations do concern processes of conception, gestation, nurturance, and birth. However, different cultures depart from standard Western biological conceptions of these processes in important ways. Many cultures draw a distinction between parallel cousins (a child of either a mother's sister or father's brother) and cross cousins (a child of a mother's brother or a father's sister), thus splitting a unified category in our system. Some Australian Aboriginal cultures have a single term for a person's father and the father's brothers, thus from our perspective lumping fathers with a species of uncle. This can give rise to sharp contrasts between consanguineal relations recognized by some culture and relations of biological kinship. For example, biological fathers might stand in no special consanguineal relation to their offspring, as in the Nayar kinship system (Gough 1959). Illustrating the opposite tendency is the partible paternity system reported for many South American tribes (Beckerman and Valentine 2002). Under this system, each of the male lovers a woman takes during pregnancy is conceived of as a father to her offspring. These men are all understood as helping to generate the offspring via the sperm they contribute (Walker, Flinn, et al. 2010).

Kinship plays a foundational role in organizing human social behavior. At the individual level, those whom we care for, live with, and marry are shaped by kin relations. These relations are normatively laden. A mother is seen as having certain duties to nurture her children in virtue of being their mother. Complex marriage norms are well known from many hunter-gather societies, and these are often defined by kinship relations, sometimes very complex ones. For example, in so-called "bifurcate merging systems" of kinship, one cannot marry a parallel cousin, while marriage to a cross cousin is ideal. This kinship system and its associated cross cousin marriage rule is quite common cross-culturally. The situation is somewhat puzzling from a biological perspective, as of course cross and parallel cousins are equally closely related to the focal individual. Kinship also shapes human social behavior at more global scales, for example by underpinning politically powerful matrilineal or patrilineal descent groups. Many prestate societies are organized around clans, and these are usually conceived as all having descended from a common (usually male) ancestor, though not always a real ancestor.

Human kinship systems are important for present purposes because kin relations bind and connect individuals. Kin relations are psychologically and motivationally potent, not least because they are so normatively laden. The quasi-biological nature of kin relations very likely matters here. Motivations of affiliation and mutual support that evolved because of the shared evolutionary interests of close biological kin are extended to culturally identified kin. The bondedness of kin helps preserve cooperative motivation and commitment to individuals whom one might interact with only occasionally in adolescence and adulthood. All other things being equal, the bond between brothers decays less rapidly than that between friends. It is no accident that organizations that rely on the loyalty of their members play endlessly on the metaphor of being a family. It is thus likely that kinship systems played a crucial role in maintaining reciprocity as the time horizon of human socioeconomic life expanded. Very probably they were also necessary for human society to reach the degree of openness characteristic of modern foragers. This connection has been thoroughly emphasized by Bernard Chapais. Indeed, Chapais argues that all that was needed to turn the closed great ape social worlds into open forager ones was an expansion of kin recognition, spearheaded by the recognition of paternal kin as pair-bonding evolved (Chapais 2008; Chapais 2013).

In our view, matters are more complex, but we agree that an expanded and more lasting set of kin affiliations was one factor. For humans recognize many more kin as kin than do great apes, and kin affiliation between humans does not require coresidence. Sisters do not have to live in the same residential group to help each other at time of need. These two factors help create and sustain links between residential groups. Marriage rules that prescribe systematic patterns of out-marriage create overlapping chains of kinship ties between bands, which in turn promote peace and alliances between bands. Moreover, once kin relations are named and become topics of conversation, language can reinforce kinship ties, refreshing memories, even when the kin themselves are absent, and this too helps stabilize peaceful links between groups, even if they come to interact less frequently. This may well have been a precondition for the mass exodus of humans out of Africa between 100 and 80 kya (Gamble 2008; Planer 2020b). Humans became able to move much more freely through the landscape because their support networks were more stable over time. These supportive relations between groups made it possible for *sapiens* to colonize forbidding environments with very limited food supplies, supporting only very small residential groups. A band of a family or two would not be stable over the long term, without support from a wider network. While small groups can penetrate harsher environments, they need social risk management. They need to be able to reconnect at times of need (Gamble 2013).

The elaborated kinship systems of ethnographic report impose steep linguistic demands. These systems feature a small set of basic kin terms: "mother" and (typically) "father," "husband," "wife." The meaning of the former pair is fixed directly by the locally distinctive understanding of how people originate and develop. The meaning of basic affinal concepts is fixed by the local understanding of the nature of marriage. Minimally, marriage is associated with certain domestic rights and duties; as we have already seen, many norms are norms about kin and the appropriate behavior between kin. The rest of the terms belonging to the system are constructed compositionally, so the terms themselves have an implicit hierarchical structure. This gives complex kin terms necessary and sufficient definitional conditions—quite an unusual semantic property for everyday words. For example, your sister is the female child of your mother and your father. Your nephew is the male child of your sibling, or of the sibling of your marriage partner. Indeed, kin terms can in general be meaningfully composed arbitrarily many times (recursively and otherwise). This is somewhat disguised in English with its comparatively impoverished kin vocabulary. Moreover, kin terms name relations holding between two or more individuals considered abstractly—e.g., a cousin is a child of a sibling of one of the parents of ego, whoever ego is—and in this regard, resemble statements of norms. As with norms, kin categories are specified independently of any particular person.

The argument of this section is that the expansion of kinship systems played an important role in stabilizing cooperation on larger temporal and social scales, and in easing tensions that inevitably arise out of sharing organized around indirect reciprocity. In turn, that expanded system relied on language, or something close to language, both through the recursive specification of all but the primary kin vocabulary, and because recognizing most kin in other residential groups depends on genealogical information that must be linguistically encoded and expressed. Knowing who your relatives are, when they live in other residential groups, depends on testimony, not first-person observation. For the most part, an agent can recognize his or her primary kin from personal memories of direct interaction. The testimony of others is not required. That changes once a kinship network is spread in physical and social space. Then the testimony of others is required for ego to identify and categorize the links in that network.

7.3.3 Stories and Gossip

Stories entertain and amuse us. These functions are important, for as Wiessner's description of bushman life shows, shared enjoyment of narrative eases social tensions and brings people together (Wiessner 2014). But stories do a lot more than this. They convey critical information about cultural practices and beliefs, including group norms (Salali, Mace, et al. 2017) but also much else about daily life (Sugiyama 2001; Smith, Schlaepfer, et al. 2017). They relate the inner workings of cultural institutions, such as marriage, exchange systems, and rites of passage. Myths are stories that concern the supernatural. Origin myths purport to explain the ultimate source of the group and its place in the cosmos, and perhaps these also encode utilitarian information in readily remembered form (Kelly 2015), as well as revealing, illustrating, and vindicating the distinctive norms of the community. Legends are stories that relate the incredible exploits of human agents, typically to illustrate a lesson or moral principle. A young person can expect to glean much of the general information they need to navigate their local social world from their group's stories.

The myths and legends of a group form part of a larger package that mark the group out as a distinctive cultural unit. This package is shared, not at the band level but rather at the level of clan or metaband. It thus helps to create and maintain social bonds between individuals living in different bands. Other elements of this package include ritual, art, and song. Taken together, these cultural elements make up the ceremonial life of the group. These elements typically interact and mesh with each other in complex ways. Howard Morphy has described and analyzed these interconnections for the Yolngu of northeastern Arnhem Land in great detail, with a special focus on myth and art (see especially Morphy 1991). Different clan groups in this region possess different sets of images or designs. These designs, and their component elements, are understood as having originated in the actions of (supernatural) Ancestral Beings to which the clan is connected. For example, Gutmatj clan designs prominently feature a particular form of linked diamond shapes. According to Gutmatj mythology, this design was first created when a bushfire swept up to an area where Ancestral Beings were performing a ceremony and burned one of their clapsticks (Morphy 1989). The design was then reproduced on the chests of these Ancestral Beings and was later passed down to ancestral Gutmatj humans. The design thus serves to connect modern-day humans with the activities of Ancestral Beings who created and bequeathed to the Gutmatj their country, while simultaneously referring to a particular event in the Ancestral past. Such designs centrally figure into ritual reenactments of mythical events, where they might be painted on people's bodies, on ritual objects (e.g., ceremonial poles), or drawn in the sand. Rituals of this kind, together with the designs they feature, are one of the primary ways by which a clan's sacred beliefs are transmitted to the next generation. The knowledge needed to correctly interpret the designs, or interpret them in a certain way (for most are multivalued), is acquired gradually over the course of life, as an individual is exposed to particular ritual performances, songs, paintings, and stories. How and when this information is to be shared with younger individuals is specified by elaborate group norms;[7] see for example Meggitt (1965).

Likewise, gossip is often in the form of narrative: "Let me tell you about what Thag got up to yesterday . . .". Gossip is central to the stability of indirect reciprocation, for that form of cooperation depends on accurate information about the social behavior of others, especially when they are out of your sight. An agent's reputation must matter, and that reputation must be accurate. The linguistic demands here are very considerable: the gossip needs to be able to specify anything from the full repertoire of the community, explain context, motivation, time, place, and consequences. It is possible to tell a story with only a protolanguage. Homesigners are convincing evidence of that. Despite having impoverished linguistic resources, they tell and understand stories (Green 2017), though this presumably depends on rich common knowledge shared between the narrator and the audience. But in the absence of full language it is very hard to imagine a group inventing a set of stories about supernatural or otherwise extraordinary events that are supposed to have happened in the distant (mythical) past; more generally, to imagine any storytelling set outside a shared experiential world. Once the scope of cooperation has expanded over space and time, reputation must be tracked over events that transcend the joint experiential world of those sharing a story. In the worlds of ethnographically known foragers, social life, including cooperation and cooperation failures, takes place on this expanded scale. Just as norms, kinship, ritual, and art became important to promoting and maintaining social bonds, so did stories, along with the craft of telling them.

Here is the situation as we see it. In the late Pleistocene, the character of human cooperation changed in important ways. Those changes made cooperation even more powerful. Those late Pleistocene humans managed risk better, securing more resources more reliably. But these changes came with costs, for they generated more potential points of social friction, and they made monitoring cooperation and keeping freeriding in check more difficult. These challenges were met by new social techniques (or perhaps by upgrading existing ones). But these in turn required the communicative resources of full language, or something close to full language.

7.4 Why Were Late Humans So Cooperative?

We last shared a common ancestor with Neanderthals between 400 and 700 kya. Something happened in the line leading to humans that put us (and possibly the Neanderthals) on course to become even more dependent on cooperation than earlier hominins. We have argued that the intensification of cooperation in our line introduced novel forms of social stress which full language evolved to help mitigate. If so, and if these changes to cooperative life were indeed unique to humans, or at least much more pronounced in humans than in Neanderthals, then our account would imply that only humans evolved full language. We must now turn to this expansion of cooperation in our line. For we have promised an account of the evolution of language without miracles, an account that does not depend on improbable coincidences or inexplicable jumps. If our picture of the evolution of language from simpler protolanguages depended on an inexplicable expansion in cooperation in the AMH lineage, we would have a miracle, or at least a mystery, buried in our theory. Sadly, we do not have a full and final answer to this question. So instead, we will survey and explain a selection of reasonable possibilities, ones which mostly complement one another. None of these possibilities are adequate as stand-alone explanations. Indeed, we discuss the first only briefly and mostly because it is the most commonly advanced. We discuss two other ideas in more detail: one suggests that AMHs cooperated more because they were less reactively, less impulsively, violent. The other suggests that AMH cooperation was a response to special environmental stresses. We think these are more promising, but they are at best incomplete. So the aim of this final section is modest: to show that there are potential explanations of the evolution of AMH super-cooperation that have some support and are worth further development.

7.4.1 Intrinsic Cognitive Differences

We begin with the most common suggestion, but one on which we place little weight. For a long time, the default assumption was that behavioral differences between anatomically modern humans and Neanderthals were due to intrinsic cognitive differences between the two species: *sapiens* were inherently "smarter." This attitude has waned with the "demographic turn" in archaeology (Powell, Shennan, et al. 2009), with its emphasis on the importance of the social context of innovation and cultural adaptation, but there are still a number of theorists who defend a view along these lines. Despite its continued support, this family of ideas faces very serious challenges. First, in some versions, the answer to "why only us" turns out to be "we got lucky." There just happened to be a cognition-enhancing mutation in the AMH lineage, but not in the Neanderthal lineage. While contingency, luck, certainly has effects on evolutionary history, the more dramatic and abrupt the mutation-triggered upgrade in cognitive powers, the less likely the scenario. We expect important differences between AMH and Neanderthal capacities to be the result of incremental divergence, and that in turn suggests a sustained difference in selective regime. Second, empirically grounded claims of cognitive difference between AMHs and Neanderthals depend mostly on inferences from subtle differences in skull shape (though there are modest but suggestive genetic differences as well). Even if differences in skull shape do map onto differences in brain size and organization, we need to be careful not to overinterpret these, as there is a good deal of variation in brain size within AMH populations and increasing evidence of a quite variable relationship between neural region and cognitive function. Our brains seem to be remarkably adaptable. All that said, there is some plausibility in the idea that there might have been intrinsic cognitive differences between AMHs and Neanderthals. The two are genetically very similar, but not identical. Moreover, at least some of the genetic differences between us and them are known to concern genes that are expressed in the brain. Hence, it is possible that the greater social complexity manifested by AMHs can in part be traced back to an intrinsic (i.e., genetically caused) cognitive difference between us.

What kind of cognitive difference? And why would it have evolved in our lineage and not the Neanderthals? One suggestion has been that *sapiens* evolved enhanced planning and/or reasoning abilities relative to Neanderthals (Coolidge and Wynn 2005; Coolidge and Wynn 2018). Another is that we evolved superior powers of social cognition (Dunbar 2009; Pearce, Stringer, and Dunbar 2013; Gamble, Gowlett, and Dunbar 2014). A cognitive upgrade of either kind would be expected to make more sophisticated forms of cooperation and coordination possible. One might see some support for one or both of these suggestions in recent neuroanatomical evidence.

Neanderthals were physically larger than *sapiens*, and yet our brains were roughly the same size. In general, a larger body requires more neurons dedicated to bodily control and upkeep than a smaller body (Striedter 2005; Herculano-Houzel 2016). Thus, it would appear that *sapiens* had a larger number of neurons available for cognitive functioning (though only if the neural density of the two species is the same). In addition, Neanderthals likely dedicated more neurons to visual perception and processing than *sapiens*. Their eyes were physically larger (as judged by larger orbits), reflecting the lower average light levels of their higher-latitude environments (Pearce, Stringer, and Dunbar 2013). Modern humans who descend from populations that have historically lived at higher latitudes are known to have larger eyes and (hence) a larger visual cortex. This is so despite the fact that they have only lived under such conditions for a relatively brief period of time, evolutionarily speaking. Neanderthals had hundreds of thousands of years to adapt. If Neanderthals had to devote more neural resources to visual processing, this suggests that our lineage had more neurons available for cognitive functioning. Endocranial evidence also suggests that Neanderthals had a smaller, flatter parietal region than *sapiens*. This brain area is known to be involved in working memory. Similar evidence suggests increased connectivity between this region and the frontal lobe, which is suggestive of enhanced executive control (Bruner 2010).

The suggestion that Neanderthals likely dedicated more neurons to visual perception and visual processing than did *sapiens* has the advantage of explaining why *sapiens* had more advanced working memory. It is not just our good luck. The high latitudes at which Neanderthals lived forced a tradeoff between visual processing and working memory that *sapiens* escaped. Even so, given the degree of interhuman variation in neuroanatomy, it is difficult to know exactly what to make of such evidence. The modest differences (at most) identified here are clearly not strong evidence of intrinsic cognitive differences, though it is certainly possible that the greater complexity of human social behavior was in part caused by a change in intrinsic cognitive capacities. It is however very difficult to see how such an idea could be developed and tested, especially given increasing evidence of plasticity in brain development. The specific social, physical, and learning environment has a powerful effect on neuroanatomy, as Celia Heyes, for one, has pointed out in arguing for her cultural evolutionary model of human cognition (Heyes 2012; Heyes and Frith 2014; Heyes 2018). So while we do not rule out this possibility, we rest nothing on it.

7.4.2 Reduced Reactive Aggression

Another prominent idea is that *sapiens* became much less prone to reactive aggression (Wrangham 2018). Reactive aggression is "hot," spontaneous, emotionally charged, and spur-of-the-moment aggression. Reactive aggression contrasts with so-called proactive aggression, the latter being "cool," calculated, and controlled, with a goal in mind. Perhaps somewhat surprisingly, there is suggestive evidence that the propensity to reactive aggression has a fossil signature (more on this presently), and that reactive and proactive aggression are controlled by different mechanisms, and hence can evolve independently of one another. For these two forms of aggression are underpinned by different neural circuits. As one would expect, the system supporting reactive aggression makes central use of limbic system structures, whereas the system supporting proactive aggression more centrally involves frontal lobe structures (Flynn 1967; Meloy 2006). This is important, for it suggests that the two forms of aggression may well be independently tunable, both in development and probably over evolutionary time as well. That is, proactive and reactive aggression are not a package deal. Recent heritability and behavioral genetics studies suggest that genes do indeed make a significant contribution to levels of proactive and reactive aggression (Segal 2012; Veroude, Zhang-James, et al. 2016). These traits would thus seem to be possible targets of natural selection.

A reduction of reactive aggression increases social tolerance, which in turn positively influences cooperation and cultural learning. Less reactively aggressive individuals can remain closer to one another, and for longer. This increases opportunities for working together, and makes joint activity less liable to fall apart. It also obviously helps with learning by observation and teaching. At the same time, lowered reactive aggression reduces within-group conflict. It does not eliminate it, for proactive within-group violence is possible, and doubtless sometimes actual. But some forms of conflict become less likely, e.g., an impulsive male lashing out at group-mates in a squabble over dividing an animal they have just killed together. A reduction in reactive aggression makes more peaceful intergroup or interband relations more likely. As mentioned earlier, bonobo intergroup meetings are not nearly as violent as those of chimps. This is very likely explained—at least in part—by the fact that male bonobos are considerably more socially tolerant than male chimps. More peaceful interband meetings build the opportunity for cooperation and cultural learning across bands. That said, the reduction of reactive aggression in late Pleistocene AMHs, if it existed at all, cannot have been very dramatic, as mid-Pleistocene hominins cooperated extensively and in ways that depended on social tolerance. The firelight niche depended on social tolerance (Planer 2020b).

The obvious question is what caused our species, and only our species, to become less prone to reactive aggression. The dominant school of thought is that we brought about this change ourselves by selecting against volatile individuals. This is the self-domestication theory of AMH origins. The attraction of this theory is its promise to simultaneously explain a range of morphological changes in later *sapiens*. Domesticated animals show a syndrome of traits. In particular, they show a tendency to retain into adulthood traits (or versions of traits) that are associated with juvenile stages in the ancestor (in the jargon, this is known as pedomorphism). In mammals, for example, the domestication syndrome includes among other things: a reduction in skull (and hence brain), jaw, and tooth size relative to the wild ancestor; a more gracile build; white fur on the forehead and feet; floppy ears and curly tails; and reduced sexual dimorphism. The dominant view historically has been that each of these features constitutes an independent change to the ancestral condition. But they seem to be a syndrome, a package deal. Using the silver fox as their model species, Alexander Belyaev and colleagues showed that the whole domestication syndrome evolves simply by selecting for reduced reactive aggression (Belyaev, Ruvinsky, and Trut 1981). Recent research strongly suggests that development of the domestication syndrome is caused by changes to the neural crest cell system (Trut et al. 2009; Simões-Costa and Bronner 2015).

According to the self-domestication theory, later *sapiens* has the (species-appropriate) signatures of domestication. In assessing this idea, we should actually compare AMHs with our ancestor, presumably African *heidelbergensis* (assuming this was indeed our last common ancestor with Neanderthals). However, the fossil record for this mid-Pleistocene species is patchy, and hence the more common comparison has been with Neanderthals (e.g., Wrangham 2018). The assumption here is that Neanderthals are anatomically less derived than is *sapiens*, especially as it became anatomically modern, relative to our last common ancestor. While this assumption is tendentious in this context, assuming what needs to be shown, it is true that if Neanderthals are used as a proxy for the immediate ancestor of *sapiens*, then we do indeed appear pedomorphic. In particular, the craniofacial and dental features of humans strongly suggest we have undergone a process of juvenilization in the last several hundred thousand years. This conclusion is further backed by recent genetic data showing strong selection in *sapiens*, but not in Neanderthals (or Denisovans), for several genes believed to be involved in domestication in dogs, cats, horses, and cattle (Theofanopoulou, Gastaldon, et al. 2017).

The earliest fossil evidence we have at present for a clear trend in the direction of *sapiens* craniofacial and dental anatomy comes from Jebel Irhoud in Morocco. These fossils are dated to around 300 kya. The Jebel Irhoud people show a notable reduction in brow ridge size, shorter faces, and certain dental similarities with AMHs (Hublin, Ben-Ncer, et al. 2017). At the same time, their brain cases do not show the characteristic globular shape of AMHs, and the face and teeth are still very large. The full suite of AMH craniofacial and dental features only becomes visible around 200 kya. Proponents of the self-domestication theory interpret this as evidence for a gradual process of self-domestication spanning 150k years or more. The self-domestication theory would be in trouble if modern craniofacial and dental anatomy had appeared early, well before 200 kya. For then there would be a puzzling time lag between the appearance of modern levels of docility, as shown by these anatomical features, and the cultural features supposedly made possible by those levels of docility and social tolerance. These are the archaeological signatures of so-called "behavioral modernity." These include an expanded toolkit, the exploitation of a broader range of resources, especially including marine resources, and the origins and expansion of material symbols in the historical record, beginning with the use of ochre (McBrearty and Brooks 2000; Henshilwood and Marean 2003; Sterelny 2011). These are collectively thought to indicate greater cultural complexity, and begin to appear in southern Africa from about 150 kya: for instance, the systematic use of marine resources dates to about 164 kya (Marean, Bar-Matthews, et al. 2007).

However, even if *sapiens* self-domesticated, thus allowing a still more cooperative and complex culture to emerge, we need an account of why only *sapiens* self-domesticated. There are different versions of the self-domestication theory on offer, with different theorists fingering different human-imposed selective pressures. In one version, selection against reactive aggression was mainly driven by female mate choice (Gleeson and Kushnick 2018). This mechanism has also been offered to explain self-domestication in bonobos. Another version of the theory treats cooperative partner selection as the main driver (Tomasello 2016). Two other versions feature a more active—and bloody—role for humans in driving the selective process. They are both versions of the so-called execution hypothesis, originally sketched by Darwin (1871), and defended in a different form by Chris Boehm (Boehm 1999; Boehm 2012). This hypothesis holds that human reactive aggression was reduced by the systematic execution of violent individuals at the hands of cooperating groups of males. Bingham (1999) emphasizes the role of weaponry in making lethal punishment possible at manageable levels of risk for within-group males, while Wrangham (2018; 2019) posits a linguistic advance that made conspiratorial plotting possible. These ideas all face problems, and none really addresses the supposed contrast between *sapiens* and Neanderthals.[8] None of them explains why these selective processes took place in the sapiens but not the Neanderthal lineage.

There is an environmental hypothesis on the table which would explain this divergence. This approach connects reduced reactive aggression to an increase in population size (Cieri, Churchill, et al. 2014).[9] The idea here is that, as population density increased, so too did selection for social tolerance, as greater tolerance was necessary to exploit resources already being exploited by other humans. Endless conflict from enforced social proximity was just too expensive. This hypothesis derives support from island-dwelling species. These typically live at much higher population densities than their mainland relatives, and they tend to be less reactively aggressive than their mainland relatives, a phenomenon known as the "island effect" (Stamps and Buechner 1985). The island-dwelling Zanzibar red colobus monkey exemplifies this in primates. These monkeys are more pedomorphic and more tolerant relative to inland red colobus monkeys (Rowson, Warren, and Ngereza 2010; Wrangham 2018).

Needless to say, this theory requires human population size to have been on the rise over the relevant time periods. This version of the theory predicts that population size was increasing before 300 kya and continued to increase until at least 200 kya (recall the gradual change in craniofacial/dental morphology mentioned above). Unfortunately, it is very difficult to test such claims. Studies comparing modern human genomes have returned conflicting results about ancient population sizes and trends.[10] What we do know is that, beginning about 800 kya, the Earth entered a period of dramatic climatic fluctuation. This period is divided into 19 different marine isotope stages (MISs). In Africa, periods of global cooling would have caused increased aridity. As baseline primary productivity must fall when the climate is cold and arid, one would expect terrestrial mammals living in Africa to undergo population contraction during periods of global cooling, all other things being equal.[11] During MIS9, which spans 336–303 kya, global temperatures were high, suggesting that human population size in Africa may well have seen an increase. However, the next 100k years (MIS8, and roughly the first half of MIS7) mark a period of significant global cooling, again predicting a consequent decrease in population density. This suggests that the trend toward increased docility and hence pedomorphism between 300 and 200 kya (if there was in fact such a trend) was caused by something other than increasing population density.

In short, some form of self-domestication may well be part of the explanation of the increased scope and complexity of late Pleistocene cooperation. There does indeed seem to be some fossil signature of self-domestication. But it is at most part of the story. For one thing, we still need a plausible environmental—or *social* environmental—factor acting on *sapiens* but not on other late Pleistocene hominins. For another, mere tolerance does not in itself result in active cooperation. As we know from actual domesticates (e.g., cows, sheep), a species can be extremely docile without positively cooperating. Docility only makes more advanced cooperation possible, rather than driving it. If self-domestication was important to our becoming super-cooperators, it is only because that process took place against a backdrop of already very advanced forms of cognition and communication.

7.4.3 Environmental Stress

So far, we have considered a change in temperament and a change in intrinsic cognitive capacity as possible triggers of the intensification of cooperation in our line. In doing so, we have been looking at agents rather than their environments. We turn now to primarily environmental explanations, using recent work by Curtis Marean (2010; 2011; 2014; 2015) as our example of such a theory. We think Marean's theory is intriguing, and in some respects very plausible. But we do not endorse it. Rather, it is an example of a class of theories that point to increased external stress on hominin life, as the late Pleistocene glacial/interglacial cycle intensified. Late Pleistocene hominins lived in perhaps excessively interesting times. So our discussion here links back to the discussion at the end of section 2.4, and the interactions between internal and external factors in driving a lineage's evolution.

As mentioned earlier, global cooling is associated with increased aridity in Africa. MIS6, which spanned 190 to 128 kya, was an unusually long glacial period, with resource zones becoming increasingly patchy and widely distributed (Mirazón Lahr and Foley 1998). That was especially true of South Africa (Marean 2011). Humans were living all over Africa through these times, but Marean has argued that MIS6 forced South African humans to make increasing use of coastal resources—in particular, intertidal shellfish. This, he plausibly suggests, transformed human cooperation and culture in a variety of ways. Intertidal taxa are densely distributed, predictable, high-quality, and available all year round. Nevertheless, as Marean persuasively argues, adapting to a coastal way of life is challenging. The return distance individuals can walk over the course of a single day defines the daily foraging radius around a hunter-gather camp. That distance is about 10 km (Kelly 2013). Unless a group can efficiently harvest marine resources, locating camp near the coast would dramatically *reduce* foraging returns, as daily foraging radius is essentially halved (one can walk 10 km inland, but not 10 km out to sea). However, to efficiently harvest intertidal taxa such as shellfish, the group must understand how the availability of those resources varies with both daily and monthly tidal patterns. Gathering intertidal resources is generally only possible during low tides, which tend to occur only once per day during daylight hours. Spring and neap tides vary according to the lunar cycle. Spring tides occur when the moon is either full or new, and hence is aligned with the sun. During these periods, high tides are at their highest and low tides at their lowest. Neap tides occur seven days after spring tides, when the sun and moon are maximally out of alignment. This difference is extremely important, for the spring low tide makes resources available that are normally hidden. In contrast, at a neap low tide, normally safe zones may become dangerous. Where the coast shelves gradually into deeper water (as is the case along the South African coast), the difference in intertidal foraging returns between spring lows and neap lows is very substantial. Hence, it is no surprise that ethnographically known people who practice a marine-based foraging economy keep meticulous track of spring and neap tides (see, e.g., Cordell 1974; Nishida, Nordi, and Alves

2006a; Nishida, Nordi, and Alves 2006b). Prior to the invention of fishing and harpoon hunting, a group would be much better served by camping farther inland during neap tides.

While learning the relationship between shellfish availability and tidal change is trivial, anticipating the tide, and in particular learning the relationship between tidal change and lunar change, is not. Tidal foragers needed to anticipate spring and neap tides from phases of the moon. No doubt mastering these relationships depended on complex communication, as well as cumulative cultural learning. Marean even claims that this signals "complex cognition characterized by fully modern working memory and executive functions" (2011, p. 434). Moreover, he hypothesizes that the Neanderthal versions of these cognitive capacities were inferior to AMH capacities, and that this prevented Neanderthals from practicing a way of life organized around the sea (see, especially, Marean 2014). We do not need to agree with these claims to agree that tidal foraging is cognitively demanding and is likely to have reorganized human lifeways, once it became a core form of subsistence.

On Marean's view, the transition to systematic use of coastal areas had a profound effect on AMH social, technological, and symbolic life. In particular, intertidal resources could be defended, and were worth defending, given their high quality and year-round availability. This would have fostered—perhaps for the first time in a long time—increased territoriality in AMHs, with small, residential groups joining forces to defend their coastal territory from outsiders. Here we might imagine something like the principle of "complementary opposition" holding (Evans-Pritchard, 1940). This principle describes a widespread form of sociopolitical organization found among African foragers. Simply put, it refers to nested political units which sometimes fight each other, but which unite when fighting outsiders. With intergroup conflict on the rise, Marean suggests group selection became important. A psychology of cooperating with one's group, but being violent toward those identified as outsiders, established. The symbolic trappings of modern forager life also established, as intergroup signaling became more important. Ritual became an important means of promoting and maintaining ties above the level of the band. We have grave doubts about much of this, but we suspect Marean is correct in thinking that coastal adaptation was a major driver of AMH social and cultural complexity over the last 200k years. It gave our ancestors access to a novel class of resources, ones that were potentially rich and (fairly) predictably located in space and time. It thus encouraged technical and social experimentation in a new ecological zone. In particular, his theory cleanly explains the fact that the earliest unambiguous markers of symbolic behavior—including the much-discussed instances of abstract art—appear at coastal sites. The theory also fits with demographic theories of cultural complexity, as coastal adaptation tends to drive up population size (Marean 2014).

Time to draw breath. This chapter has focused on identifying the selective forces that drove the evolution of an increasingly rich set of protolanguages. In our view, that shift to increasing linguistic complexity was a response to the changing character of human cooperation, and hence to the changing cultural tools needed to manage that cooperation. These required storytelling, a shared and rich ritual life, gossip, and explicit norms. In turn, these all require richer expressive powers, including an expanded repertoire of speech acts, an expansion of mood, a much richer lexicon to describe the imaginary and the supernatural, and an ability to exactly specify the acts and motivations of other agents, both in gossip and in contested cases of norm violation. Moreover, this multifaceted expansion of expressive power took place in a social context in which one could rely less on common knowledge. Indeed, in part this is why selection favored an expansion of expressive powers. These different vectors of expressive power all come in degrees. We have not attempted to specify the extent to which they all expanded in synchrony, or to give a fine-grained treatment of the incremental expansion in each vector. We have not, for example, attempted to identify the stages through which storytelling became more elaborate and less restricted to the mundane. We doubt that the evidence as we now have it would support a more fine-grained scenario of this expansion of expressive capacity. Given that, our account is really a sketch or an approximation of a lineage explanation of the transition from a quite rich but mundane and utilitarian protolanguage to full language.

This coevolution of social and linguistic complexity is founded on the idea that the character of human cooperation changed in important ways. For this reason, in this section we have considered several factors that might explain the fact that *sapiens* found themselves on a trajectory toward hyper-prosociality, and hence full language. Our aim has not been to provide an exhaustive survey of the options, still less to endorse one. Indeed, we strongly suspect that the *sapiens* expansion and transformation of cooperation, leading morphologically to AMHs and socially to "behaviorally modern" humans, depended on the simultaneous effects of a number of factors, and the extent to which it was unique remains open. However, we think we have done enough to show that in proposing that cooperation transformed and expanded in the late Pleistocene, with profound consequences for language, we have not tacitly relied on a miracle, an inexplicable jump in social complexity.

8 Without Miracles?[1]

In section 1.1, we set out criteria that should be met by any credible account of the evolution of language. To recapitulate, we suggested (following Kevin Laland, in particular Laland 2017):

1. A credible theory of the evolution of language should be gradualist and incremental, specifying a sequence of changes in capacity from baseline agents to agents with capacities roughly equivalent to those of living humans.
2. The choice of baseline capacities must be principled, supported by independent evidence.
3. Each step in the sequence of changes should be fairly small, and plausible both as a minor change in the proximate mechanisms and/or the social environment of the agents in question, and plausibly advantageous, given what can be inferred about the selective environments of those hominins. The account should not depend on very improbable events, or improbable selective regimes.
4. The account should be congruent with the dynamism and variability of hominin environments at and over time.
5. While some elements of speculative reconstruction are probably inevitable, the theory should have testable implications about the historical record. The more a theory does this, the more serious attention it deserves.

Any account satisfying these criteria relies only on well-understood, nonmiraculous evolutionary mechanisms. In addition, and more specifically, the theory should explain:

6. The honesty of language, and hence why senders keep sending, and receivers keep attending.

7. The distinctive scope and expressive power of human language.
8. The uniqueness of human language: why it evolved only in the hominin lineage; why, perhaps, full language evolved only in the *sapiens* twig of that lineage.

It is time for a self-assessment. We make no claim that the account of the emergence of language developed in this book fully meets these criteria. For there are important aspects of language about which we have said little. We have not offered a detailed account of dual coding: the emergence of phonological and syllabic structure in lexical items. While we have offered an incremental account of the evolution of the critical cognitive capacities on which syntax depends, we have not given an incremental account of the evolution and elaboration of the many syntactic devices exploited in the hierarchical organization of living languages. In discussing the cognitive foundations of hominin communication, we have, with the exception of chapter 5, mainly confined ourselves to describing capacities—Dennett's "performance specs" of hominin minds—rather than specific cognitive architectures. Our account of the expansion of the lexicon is very basic: we do not discuss the acquisition of modifiers like adverbs, the distinctions between mass and count nouns, or between properties and quantities. Many of the sophisticated features of living languages are out of our reach. So at best our solution is partial. That said, the model we develop satisfies these criteria in important ways.

The baseline. We offer a principled account of the baseline capacities of early hominins, mostly grounded on comparative considerations about primate abilities, but also on what is known of the lifeways of late Pliocene hominins. Despite the complications of recent data, we think those comparative considerations still support the idea that quite minor social and cognitive changes would have sufficed for a much richer system of gesture, whereas much more dramatic changes would be needed to harness vocalization as a medium of richer communication for Pliocene and early Pleistocene hominins.

The initial expansion of communicative abilities. Our account of the early expansion of expressive capacities from that baseline is both incrementally plausible and empirically principled. It is incrementally plausible because the cognitive and social capacities it presupposes are only modestly more sophisticated than those actually at the disposal of great apes. It is empirically principled, for the historical record of the lifeways of late Pliocene and early Pleistocene hominins suggests that their capacities for causal and social reasoning exceeded those of living great apes.

Gradualism. More generally, our overall account of the evolution of the capacity for language and its predecessors is incrementally plausible, for it does not rely on the emergence of entirely novel cognitive capacities. In our view, most and perhaps all of the cognitive capacities on which language depends are elaborations of capacities present in a more rudimentary form in great apes, and hence, very likely, in early hominins. That supports testability, for if the cognitive capacities that support language support other forms of behavior as well, we expect them to leave more direct historical traces.

Displaced reference. Likewise, our account of the emergence of one of the most salient features of language, displaced reference, is both incrementally plausible and empirically principled. We take the abilities required for displaced reference to consist of: (i) the ability to plan and launch a precise motor sequence independently of its normal environmental trigger (control by recall rather than recognition); (ii) improved social reasoning: the ability to reason about the probable reactions of others to sign sequences; (iii) more elaborate communicative intentions, not just getting another agent to do something here, now; (iv) the ability to recognize and plan for anticipated future needs. We argue that information about erectine lifeways supports the view that they had these capacities. It is one thing to have the potential to communicate with displaced reference; another to realize that potential. But communication with displaced reference is adaptive if response to future needs is sometimes collective rather than individual, and if planning is at all public rather than fully private. Once again, we argue that information about erectine and heidelbergensian lifeways makes that likely (though perhaps not certain).

An expandable lexicon. The expressive power of language depends in part on an expandable lexicon. Our account of such a lexicon is incremental in two respects. In contemporary languages, expansion is made easy by a metalinguistic apparatus embedded within the languages themselves: we have expressions for different kinds of words (Sterelny 2012a; Sterelny 2021). Moreover, living humans have very sophisticated social reasoning abilities, and that makes it possible to interpret novel signs invented on the fly.

We take it that everyone agrees that these are late-arriving features of language and mental life. But we show that a sign system is quite expandable (i) if the hominins using it are willing to experiment (and to try to interpret others' experiments), (ii) if those hominins have causal and social reasoning abilities significantly superior to those of great apes, (iii) even if those reasoning abilities are much less powerful than those of living humans. As before, information about Pleistocene lifeways supports the view that hominins of that time were better at causal and social reasoning than great apes, and it also supports the view that an expandable lexicon would confer on them a fitness advantage, especially in guiding social learning about their local patches.

In our view, iconicity was important in the early expansion of hominin sign systems, and iconicity may well have remained important as a developmental scaffold for juveniles through much of the Pleistocene. But if all wordlike signs were icons, iconicity would impose very considerable constraints on possible lexicons. If mid-Pleistocene foraging depended on natural history competences analogous to those of living foragers, those mid-Pleistocene foragers could recognize, think about, and probably needed to refer to many kinds, places, and properties for which there would be no suitable icon. If so, mid-Pleistocene hominins needed to be able to add arbitrary signs to their lexicon, if their cognitive and social capacities made that possible. Here our rejection of the index-icon-symbol framework becomes important. In our view, the transition from iconic signs to so-called "true symbols" crosses no cognitive divide. Iconicity comes in degrees, and some chimp gestures are so abbreviated as to lose their iconic properties. Moreover Kanzi mastered arbitrary signs. The cognitive demands here are on semantic memory and social intelligence. So in our view, while iconicity helped establish and stabilize mutual signaling as an important feature of Pleistocene social life, no novel capacities were needed to gradually incorporate less iconic and noniconic signs into their repertoire.

Structured signs. Obviously the expressive power of language and its predecessors also partially depends on structured signs, and ultimately hierarchically structured signs. Our incremental account of the emergence and elaboration of structure depends on four ideas. (i) A gesture-first account of the evolution of language finesses the problem of explaining how structured signs evolve from holistic signs. Gestural signs are almost primitively structured. While the very simplest gestures—a point, an attention-getter—are not structured, any elaboration introduces structure. (ii) Evidence from baboons (especially) and other primates shows that receivers are preadapted to extract information from sign sequences. If the sign sequence carries information, receivers can extract it. More speculatively, but in our view very plausibly, Pleistocene hominins were, we think, trackway readers, and that further developed their ability to recognize information implicit in sequences. (iii) Our picture depends on a specific and admittedly speculative hypothesis about baseline structured signs: they consisted of a point plus some further gesture (or gestures) that elaborated or disambiguated the point. In our view, while unproven, this is empirically plausible. For great ape gesturing is almost at the threshold of such simply structured point-gesture combinations. (iv) Finally, we argue that competence with hierarchically structured complex signs depends on the same cognitive capacities that are manifest in mid-Pleistocene lithic technology. In particular, the capacity for the hierarchical control of complex motor sequences is manifest in the developed Acheulian and successor technologies. The capacity to read and parse such sequences is shown in the social learning of these technical skills (at least if that learning depends on imitation). These cognitive abilities made the use of hierarchically structured signs possible. The use of such sequences became advantageous, we argue, as changes in the late Pleistocene social worlds eroded the extent of common knowledge, of information more or less automatically shared by everyone in the conversation circle in virtue of living and acting together. Such shared information could be left as implicit background. As that common pool shrank and information gradients increased, more had to be made explicit, and syntactic organization is an efficient means to this end.

Speech act richness. Our account of the functional expansion of protolanguage—of the emergence of a richer set of speech acts—is tied to our view of late Pleistocene changes in social life, rather than changes in the intrinsic cognitive capacities of Pleistocene hominins. Such intrinsic changes might well be important too: perhaps the evolution of norm psychology; perhaps the evolution of an imagination that can understand and enjoy stories as stories. If so, our account of the emergence of a richer menu of speech acts is partial rather than complete. As it stands, our account focuses on changes in the character and scale of cooperation late in the Pleistocene.

These changes were built on the foundations of earlier, simpler, and more limited forms of cooperation, and they made cooperation more pervasive and profitable, but also introduced new potentials for conflict. We outlined a set of cultural innovations with the potential to damp down conflict and reduce its costs. These cultural innovations—a greater role for explicit norms, shared rituals, a shared narrative life, rich and accurate gossip—can all exist in rudimentary or more developed forms. Even in contemporary life, many norms are implicit. But in their more developed forms, all these innovations presuppose language capacities close to those of living humans, the ability to do many different things with words: joke, tease, reprove, remonstrate, praise, entertain, inspire, arouse, terrify.

From gesturing to talking. A gesture-first theory of language, like ours, faces an additional burden. We need to give an empirically plausible, incremental account of the shift from a mostly gestural system to a predominantly vocal one. We attempt to meet this challenge through our analysis of the firelight niche. We think this niche became important to humans perhaps in the period 1 mya–800 kya. One of its effects was to expand the length of the usable day; another was to encourage (perhaps even enforce) longer periods of very close association between group members. We think these changes mediated the transition to a mostly vocal channel. Our account of this change depends on three linked ideas: (i) We think there was a gradual increase in control over vocalization, driven mostly by selection for proto-song (though selection for the vocal imitation of animals, important to many foragers, may also have played a role) and perhaps selection for laughter. Here we depend on the ideas of Robin Dunbar. He shows that the stress on time budgets increased as social life became more complex, and pointed to the greater efficiency of song and laughter (as contrasted with grooming) as forms of stress release and conflict reduction. The firelight niche particularly demanded the control of conflict, as it enforced close proximity and made managing stress by spatial separation dangerous. (It is not safe for Thag to wander off into the dark because he is annoyed with Thug.) (ii) Once the vocal channel is available, it offers various efficiencies over gesture. (iii) Firelight is dim, so the firelight niche was a suboptimal site for gestural communication. But it is friendly to vocal communication, if the party is packed compactly around a fire (or between two: many Australian Aboriginals sleep between two small fires). In these circumstances, a mixed-modality system can readily shift to one primarily relying on the vocal channel.

Cultural learning. In addition to much else, the evolution of rich protolanguage and language itself depends on the parallel evolution of greatly enhanced capacities for social or cultural learning. As we explained in section 1.5, protolanguages are difficult cultural learning targets. Once the lexicon has expanded, the informational target is large. It is error-intolerant, as small differences in signal form often make a difference to meaning. Language in use is fast and ephemeral, so there is no opportunity to learn by emulation. The data are often messy, as many utterances leave much of what is meant implicit, and there are many potential ambiguities. It is true that as the capacity to communicate became critical to a hominin's life prospects, that would have selected for improved social learning. However, in our view, the most important initial factor in selecting for improved social learning was the increasing informational demands of hominin foraging. There is some debate about the importance of social learning in the earlier stages of Pleistocene stone toolmaking, with some claiming that we have a clear signal of social learning only with the emergence of Levallois flakes and microliths (Tennie, Call, et al. 2009; Corbey, Jagich, et al. 2016; Tennie, Braun, et al. 2016). In our view, as we explain in chapter 5, this much understates the cognitive challenges of developed Acheulian technology. It is striking that even with expert tutoring, it takes living humans years to master these skills. But even if we are wrong about this, if we are right about early and mid-Pleistocene hunting, that in itself is a social learning signature. For hunting, and much gathering, depend on rich and accurate natural history information about one's local patch, and a hominin would die of starvation before mastering this information on their own. As Peter Richerson and Rob Boyd point out, eighteenth- and nineteenth-century explorers did routinely die of starvation because they could not learn for themselves crucial information about local resources in time (Richerson and Boyd 2005; for a particularly forlorn example, see Murgatroyd 2002).

Great apes are quite good social learners, because they are quite good learners in general, and that makes it possible for them to do a good deal of social learning. We think we can safely assume that early to mid-Pleistocene hominins were even better. The Pliocene and Pleistocene archaeological record needs to be read with much caution, for it is biased. All sites degrade and disappear; the further back they are made in time, the greater the likelihood that they will have vanished. Likewise, very little of the material culture of a group survives. All else equal, the items that are made most frequently are most likely to leave traces. Simple items of kit are made much more frequently than complex ones, because they are cheaper. Combining these two biases, the material traces of ancient hominins is likely to underrepresent their material culture, especially with respect to the more complex elements of that culture. The more ancient the hominins, the greater this underrepresentation. However, despite these cautions, we think the archaeological record supports the view that social learning was becoming more reliable and more efficient from 500 kya, and especially from about 200 kya. There seems to be a signal of technological and ecological innovation accelerating, and of toolkits becoming richer and with greater regional specificity. This supports the idea that from about 200 kya on, social learning capacities were advanced enough to support the cultural transmission of very rich protolanguages and languages.

As with most of what we have said about the evolution of language, in our view the individual cognitive capacities that support reliable and efficient social learning mostly exist in great apes and existed in very early hominins, though in much more rudimentary forms. These include decent semantic memory; imitation learning (though not routinely used); emulation learning; causal reasoning (perhaps very rudimentary in great apes); some theory-of-mind capacities. The exceptions might be vocal imitation; episodic memory; active teaching.

Obviously, socially learning a vocal language is impossible without vocal imitation, and very likely early hominins could not vocally imitate. We suspect, but cannot prove, that vocal imitation required only bringing voice under fine-grained executive control (for once they tried sounding protowords out, correcting feedback could come from the group). For Cheney and Seyfarth's work with baboons suggests that early hominins attended to and could remember the vocal sequences of other agents. What of episodic memory? With episodic memory, living humans do not just remember specific episodes (what, where, when, whom). In addition to recalling what happened, they recall the getting of that information. Episodic memory is often autobiographical, reliving by "mental time travel" intense moments of one's life. But it need not be: for Sterelny, but not Planer, the assassination of John F. Kennedy is an episodic memory, as he recalls the news being announced on a school excursion. It is very unclear whether great apes (or any other animals) have this form of memory. But it is also unclear whether it leverages social learning. Teaching, in contrast, is critical to human social learning (this was once controversial, but is now widely accepted). As we see it, the emergence of teaching rested on (a) motivational changes: a general increase in prosociality and (b) an improved theory of mind, especially becoming sensitive to the value of information as a resource, in ways great apes are not sensitive. We have argued that the archaeological record supports the view that mid-Pleistocene hominins had become more prosocial and had enhanced theory-of-mind abilities, in comparison to baseline hominins. These changes opened the door to teaching, and we accept Peter Hiscock's argument that learning stone tool skills would strongly select for opening that door.

In short, we think we have given an incremental and archaeologically grounded account of building the social learning capacities that made advanced social learning possible, though that account is presented in (much) more detail elsewhere (Sterelny 2012a). In addition, we think the social environment changed in ways that supported more reliable social learning. Those changes included (i) a more socially tolerant environment; (ii) reproductive cooperation, giving young children access to a greater range of models; and (iii) more open and tolerant relations between residential groups, buffering them against accidental loss of expertise, making it possible to borrow innovations from others, and giving near-experts access to a wider group of skilled models.

Honesty and uniqueness. Conversation is not just information transfer. When it is, it is not always honest, and even when honest, not always reliable. Much of what others know "ain't so," as Sam Clemens wisely observed. But it is predominantly honest and often reliable, and that has quite often been seen as puzzling, given that individuals have differing fitness interests. Indeed, Dessalles's whole theory of language is organized around this supposed mystery (Dessalles 2007). We think the honesty issue is somewhat overblown, in part because modeling wrongly treats one-on-one private communication as the base case, and in part because we see honest talk as just a special case of cooperation. Indeed, in one respect it is a little less puzzling than cooperation over material resources, for information shared is not information lost (though sometimes its practical value may be diminished). Perhaps for this reason, living humans do not seem to have an irrational form of greed over informational resources as they often do over material resources, where sacrificing immediate gain for a larger but longer-term gain is often effortful and subject to weakness-of-will failures.

For us, then, the honesty and uniqueness questions become: Why did hominins evolve such cooperation-dependent lifeways? Why did only hominins develop such cooperation-dependent lifeways? Why (perhaps) did only *sapiens* develop the complex forms of cooperation that required the resources of full language? We have not answered these questions in this book, though we have said a little about the third. Rather, we have embedded our account of the evolution of language within the framework of a broader picture of the evolution of hominin cooperation. We have elaborated and defended that framework, but not here, as it requires book-length treatment in its own right (Sterelny 2012a; Sterelny 2021). If that account of the emergence and changing character of hominin cooperation is seriously wrong, this account of language falls with it. If, on the other hand, it is broadly correct, it answers the honesty and uniqueness questions.

Putting all this together then: we do not claim to have provided even a close approximation of a proper lineage explanation, taking us from an independently supported baseline identifying the communicative skills of the earliest hominins to language-equipped modern humans. But we do claim to have outlined, and in places done a little more than outline, important elements of such an explanation: most notably, an expandable lexicon, displaced reference, the core cognitive capacities on which syntax depends, the gesture-speech transition (assuming there was one), and the expanded functionality of language.

Glossary

Acheulian. A technological industry most closely associated with *Homo erectus*. It features the famous handaxe and other large cutting tools. Acheulian tools, especially later ones, are notable for their improved craftsmanship relative to Oldowan tools.

Anatomically modern humans (AMHs). A grade of *Homo sapiens* defined by anatomical dimensions within the normal range of variation of contemporary *sapiens*. To say that a hominin is a member of *H. sapiens* is not quite the same as saying it is anatomically modern: every hominin in our lineage after the split with the Neanderthals is a member of *sapiens*, but not all of them are AMHs (this distinction is important in sections 7.4.2 and 7.4.3).

Anthropomorphic fire. Fire produced or maintained by hominins, as contrasted with naturally occurring fire.

Australopithecines. Members of a (presumed) hominin genus that evolved after our line split from *Pan* but before the evolution of *Homo*, characterized by notably smaller brains and bodies than later hominins. (Some populations recognized as australopithecines lived after the first fossil species recognized as member of the *Homo* genus. So while, on the standard view, *Homo* evolved out of the australopithecines, it did not replace the australopithecines.)

Band. A stable but fluid unit of hunter-gather social organization, typically composed of several nuclear families, sometimes with other relatives. It is the unit of daily interaction, with most of its members camping together most nights.

Behaviorally modern humans. Forager communities known from the archaeological record whose social and subsistence organization seems to fall within the range of forager variation known from ethnography. Communities earlier than 100 kya are not clearly behaviorally modern.

"Bird's-eye view" representation. A representation of a social interaction in terms of abstract roles rather than specific identities of agents. Facilitates role-reversal social learning.

Coastal adaptation. A subsistence mode that revolves around marine resources. Notable for the knowledge it implies of daily and monthly tidal patterns.

Composite sign. A sign whose meaning is a function of its parts, but whose meaning cannot be reduced to the mere conjunction of the meanings of its parts.

Composite tools. Multipart tools often blending different media (e.g., stone and wood). Late middle and late Pleistocene composite tools were often hafted using heat-treated pitch (glue).

Context-free grammar. A formal grammar characterized by more complex production rules than a regular grammar can feature. Notable for permitting the embedding of one constituent inside of another (center-embedding).

Cue/signal distinction. A cue is a trait or act that makes information available to second or third parties, but which has not originated, nor been maintained, because it does so. A signal's function is to communicatively influence second parties, though not necessarily by making information available to them.

Cultural evolution. The accumulation of cultural items (construed broadly) over time via processes of social learning.

Denisovans. A sister species of Neanderthals. Fossil evidence for this species is extremely limited at present (a pinky bone and three teeth for one individual; a lower jaw bone for another).

Displaced reference. Reference to an object or event elsewhere in space and/or time.

Elaboration. A primatological term for the production of alternative gestures in the wake of a communication failure or partial failure.

Emulation. A form of social learning involving attention to an action's effects, as opposed to the action itself or the goal behind it.

Executive control. A domain-general cognitive faculty tasked with selecting, sustaining, and organizing mental representations (construed broadly).

First- and higher-order intentional systems. A first-order intentional system has beliefs and preferences, but not beliefs and preferences about beliefs and preferences (or similar cognitive states). A second-order intentional system has beliefs and preferences, including beliefs and/or preferences about beliefs and preferences. A third-order intentional system has beliefs and preferences, including beliefs and/or preferences about beliefs and preferences, when those beliefs or preferences are about beliefs or preferences. Thus Thag is a third-order intentional system if Thag recognizes that Thug believes that Thag wants Thug dead. And so on.

Formal grammar. A finite system of rules and symbols that generates a specific set of symbol strings. This set of strings may be infinite.

FOXP2. A gene (minimally) linked to fine-grained control of our vocal apparatus.

Genetic accommodation. Natural selection for genes that facilitate the development of an advantageous trait which was originally learned or acquired during an individual lifetime.

Genotype. The specific complement of genes carried by an organism. This complement of genes plays a central role in the developmental sequence leading to that organism's physical and behavioral characteristics, known as its phenotype, though there is much controversy as to how, exactly, to best characterize that role.

Grammar. The specification of the morphology and syntax (morphosyntax) of a communication system.

Great apes. The clade encompassing orangutans, gorillas, humans, chimpanzees, and bonobos.

Hafting. The attachment of a tool to a handle (haft).

Holocene. A geological epoch spanning from 12 kya to the present.

Homesign. A gestural communication system created by a deaf individual in the absence of exposure to a conventional language.

Hominin. A member of the lineage including living humans, which began when that lineage diverged from the lineage leading to the two living chimpanzee species.

Homo. The genus encompassing *Homo habilis* and all its presumptive descendent hominin species.

Homo erectus. A species of *Homo* that lived in the early and middle Pleistocene. Notable for having bigger brains and bodies than early hominins; also for having more sophisticated technology and hunting.

Homo habilis. An early *Homo* species, once thought to be the first hominin toolmakers. The presumed direct ancestor of *Homo erectus*.

Homo heidelbergensis. A presumed descendent of *Homo erectus*. Still larger brained, and with more complex technology, hunting, and social lives than erectines. The presumed last common ancestor of *Homo sapiens*, Neanderthals, and Denisovans.

Homo sapiens. A species of *Homo* that began in Africa in the middle Pleistocene, which includes all modern humans.

Hyoid bone. The bone supporting the larynx, which in turn supports the vocal chords. The structure of the hyoid bone carries information about the condition of the larynx and vocal chords.

Imitation. A form of social learning involving the reproduction of motor behavior, especially a motor sequence, and (according to some theorists) recognition of the goal behind that behavior.

Intentionality. The property of "aboutness." Thus, for example, the name "Venus" is about a particular planet; "tigers are fierce" is about the fact that tigers are fierce.

Intentionality is the defining feature of representations, including but not limited to mental representations.

Intentional communication. Communication motivated by intentions (belief-desire pairs). In the primatological literature, this is operationalized as communication involving social use, sensitivity to audience attention, and more.

Kya. Thousands of years ago.

Levallois. A technological industry associated with extensive platform preparation. A tool of the desired size and shape is struck from the prepared platform with a final blow of the hammerstone. More physically and cognitively demanding than Acheulian techniques.

Lexicon. The set of basic meaningful elements in a communication system. Words and bound morphemes in human language.

Life history. The rate of maturation of individuals of a given species. The main fossil evidence of life history relates to development of the teeth.

Lineage explanation. A series of small, incremental steps that take us from a baseline condition (e.g., in the last common ancestor of two species) to some later condition of theoretical interest. Each step in the series is presumed to be advantageous to bearers of the corresponding version of the trait, or at least not deleterious, and each step is small enough to be a single evolutionary change.

Metaband. A higher-level unit of hunter-gather social organization, composed of multiple bands. A metaband often forms an ethnolinguistic unit (i.e., a group that shares a unique language).

Miocene. A geological epoch spanning 23.03–5.33 mya.

Morphology. Principles governing the composition of basic meaningful elements (morphemes) to form words.

Mutualism. A form of cooperation in which both donor(s) and recipient(s) benefit without donor(s) paying an up-front cost.

Mya. Millions of years ago.

Neanderthals. A middle and late Pleistocene *Homo* species that has historically served as a contrast class to *Homo sapiens*. One of (minimally) two sister taxa of *sapiens*; Denisovans are another. We are estimated to have shared a last common ancestor (*Homo heidelbergensis*) with Neanderthals between 400 and 700 kya.

Norms. Emotionally charged expectations as to how individuals (including oneself) will behave.

Oldowan. A technological industry associated most closely with *Homo habilis*. The main products were sharp flakes struck from a cobble; the cobbles were probably sometimes used as hammers. Notably less sophisticated than the Acheulian. To the extent that there are reoccurring design motifs, these may be explained by natural fracture properties of the stone being worked.

Offline processing. Entertaining/processing a representation of a past or hypothetical state of affairs. Such cognition need not result in action of any kind.

Online processing. Entertaining/processing a representation of a concurrent state of affairs, typically action-oriented.

Ostensive-inferential communication. Communication involving the expression and recognition of communicative intentions. A communicative intention is an intention for one's audience to believe that one intends them to believe some proposition or to perform some act, with the audience recognizing you have that intention.

Paleolithic. The period of time during which *Homo* species have been manufacturing stone tools. Divided into Lower (~3 mya–300 kya), Middle (300–50 kya), and Upper (50–12 kya).

Pan**.** The genus encompassing chimpanzees and bonobos.

Persistence. A primatological term for the repetition of a gesture in the wake of a communication failure or partial failure.

Pidgin. An ad hoc communication system created to bridge the communicative gap between two cultures lacking a common natural language. It mixes elements of the natural language of each and commonly features a large, overlapping lexicon with a limited role for word order in encoding meaning.

Phenotype. See **genotype**.

Phenotypic plasticity. The capacity of organisms to adapt their phenotype to their environment over the course of their lifetimes. Comes in degrees.

Pleistocene. A geological epoch beginning at about 2.58 mya and closing about 12 kya. Late in this epoch, the large-brained hominins evolved. Almost everyone agrees that most of the important steps in language evolution took place in this epoch.

Pliocene. A geological epoch beginning at about 5.33 mya and closing about 2.58 mya. Stone tools were being made by late in this epoch; likewise, by late in this epoch (but perhaps earlier) some hominins were habitually bipedal.

Phonology. Principles governing the composition of sounds to form lexical items.

Protolanguage. A precursor to language as we now know it with some but not all of the design features of the latter: in particular, without the recognizable syntax or morphology of standard languages. Conceived of in this book as a type of communication system with, in its more advanced forms, a large, expandable lexicon and supporting displaced reference, but with limited grammatical structure.

Reactive aggression. Aggression fueled by emotion and prompted by immediate circumstances as opposed to being preplanned.

Reciprocity. A form of cooperation in which donor(s) incur a cost now in the expectation that it will be recouped in the future. Also known as *reciprocal altruism.*

Recursion. A property of a derivation using a formal grammar, strictly speaking. A derivation is recursive when it makes use of the same production rule more than once. This is how an infinitude of symbol strings can be generated by finite means. The term "recursion" is often reserved to refer to center-embedding, however.

Regular grammar. The simplest kind of formal grammar. A regular grammar permits one to grow a symbol string only from its edge outward. No regular grammar permits center-embedding.

Reproductive cooperation. Communal care of and/or investment in relatives, especially when they are children. Often a form of reciprocity, though considerations of relatedness are typically also relevant.

Self-domestication. The process by which a species imposes selection pressure on itself that leads to domestication, the hallmark of which is reduced reactive aggression. Domestication is linked to a syndrome of anatomical traits, some of which leave fossil evidence.

Sexual dimorphism. The degree of difference in a trait between the males and females of a species.

Social learning. Learning aided by observation or interaction with others.

Syntax. Principles governing the combination of words to form constituents and sentences.

System 1 and System 2 cognition. System 1 cognitive processes, for example face recognition, are fast, apparently effortless, readily compatible with carrying out other tasks in parallel, often automatic in the sense that the agent does not need to decide to recognize the face, and often opaque to introspection. System 2 cognitive processes (solving a logic problem, filling in a tax return) are slower, effortful, consciously initiated, less opaque to introspection, and not usually compatible with doing other tasks in parallel.

Theory of mind/mindreading. The capacity to recognize that other agents (typically conspecifics) have minds. With this capacity, in most circumstances we are reasonably good at recognizing what other humans are thinking and feeling. The extent to which we share this capacity with great apes is unclear and controversial.

Vocal imitation. The copying of (typically arbitrary) vocal behavior of others.

Notes

Chapter 1

1. Morten Christiansen and Nick Chater have underscored these processing demands and their implications for language evolution: see Christiansen and Chater (2016a); Christiansen and Chater (2016b). Stephen Levinson with various colleagues has shown how finely tuned and time-pressured the interactions between agents are: he reviews some of this work in Levinson (2016).

2. See for example Tomasello (2008); Tomasello (2014); Scott-Phillips (2015). Much of this builds on Sperber and Wilson (1986).

3. This might require some qualification. Daniel Everett has argued that speakers of Pirahã make little or no use of displaced reference, though his analysis suggests that this is more of a cultural norm than a cognitive incapacity. In any case, his analysis is controversial: see Everett (2005); Everett (2009); Nevins, Pesetsky, et al. (2009).

4. In fact this menu varies in existing languages, for as Austin (1962) points out, some speech acts depend on specific cultural institutions: Ron can become married by saying "I do" only in cultures with that form of wedding ceremony.

5. So one call might contain the sound sequence 'atemalpo'. This does indeed contain 'ma', but it might equally well be bracketed [atem] [alpo]. Without something like a division into a tractable and consistent syllabic structure, these infixes may well be cryptic.

6. For example, in the paleoanthropological literature, the appearance of material symbols (incised ochre, mortuary practices, jewelry) is often taken to mark a cognitive breakthrough of some kind, even though the supposed symbolic practices in question are decidedly heterogeneous: see for example Pettit (1993); Pettit (2011); Berwick and Chomsky (2016); Gamble, Gowlett, and Dunbar (2014).

7. Peirceans view the situation in a different and much more complicated way. The idea that icons are the simplest signs to use is rooted in Peirce's hierarchical theory of reference. On this theory, we acquire an icon for some content when we recognize

that a present experience (of a sign) resembles a past one (of its content). Hence, a picture of a dog is an icon of a dog for some agent because the agent recognizes her perceptual experience of the picture as resembling a past experience of a dog. We acquire an index for some content only when we recognize that our present experience linking two objects (the potential sign and content) resembles past experiences in which these same two things were linked. The acquisition of an index is thus seen as a more complex event, one that presupposes facility with iconic reference. The recognition of an indexical relationship between a sign and its content is based on the recognition of an iconic relationship holding among a certain set of experiences. See Deacon (1997) for a detailed exposition of Peirce's hierarchical theory of reference. Deacon goes to Herculean lengths to defend such a theory.

8. Also see Hiscock (2014).

9. The receiver need not represent the two as similar for the similarity between them to play a causal role in eliciting a functional response: the partial overlap between threat and actual preparation for attack associatively triggers an alert.

10. A cue is an act or feature of an agent that makes useful information available to other agents, but which is not produced to send information. Flight from a predator can tell others of danger, but the flight is not a signal, it is an attempt to escape.

11. Some photographs may be an exception, though as digital processing becomes more important, one needs to know how to read them too; think, for example, of the beautiful deep space photographs taken by the Hubble telescope.

12. Before the 1970s, the ethological tradition focused on apparently cooperative communication with common interests: chicks signaling to their parents and the like. This assumption of common interest and honest communication came under fire in Krebs and Dawkins (1984).

13. The game Battle of the Sexes is a famous example of such a model. One player prefers that both attend the opera; the other, that both attend the boxing match. But both prefer going somewhere together as opposed to going it alone. This game is about coordinating acts (going to the opera, going to the boxing match), not adapting a behavior to an exogenously chosen state of the world.

14. For more on the distinction between on-the-ground models and the broader sender-receiver framework, see Planer and Godfrey-Smith (forthcoming).

15. See Kuhn (2020) for a probing synthesis of the Paleolithic record of stone toolmaking as a whole, including this period. For specific discussion of blades and hafting around 500 kya, see Wilkins and Chazan (2012); Wilkins, Schoville, et al. (2012).

Chapter 2

1. Seyfarth and Cheney think that this social intelligence is conceptualized. A baboon's map of the social world is objective rather than egocentric, as it maps relations between third parties. Moreover, the social facts to which baboons are clearly adaptively sensitive—kinship organization and rank organization—cannot be characterized in sensory terms. The members of a matriline (they claim) do not share a distinctive sensory property or cluster of sensory properties. Likewise, third-party rank relations have no sensory signature. The view that social understanding is conceptualized seems plausible in light of the fact that baboons robustly track individual identity and social rank. They track identity through face recognition as well as voice recognition. Rank can be inferred from both aggression of dominants to those lower-ranked, and submission of those lower-ranked to those higher-ranked.

2. In this species, all the members of one matriline rank above, or below, all the members of any other matriline; there is an order of matriline dominance which structures all between-matriline interactions.

3. Godfrey-Smith, in his commentary on Cheney and Seyfarth (2018b), makes a similar point (p. 111).

4. See Bar-On and Moore (2017) for a critical discussion of this idea. They draw a distinction between *Carnapian* and *Gricean pragmatics*. The former just involves context-sensitive interpretation of a sign. The latter involves attributing a communicative intention, conceived of as a higher-order intentional state, to an actor. Bar-On and Moore read the baboon evidence as revealing standing pragmatic competence in primates in the Carnapian sense only, and consequently see it as at best "indirectly" connected to the evolution of language. Suffice it to say that we regard baboons' fluency in Carnapian pragmatics as much more relevant to language evolution than Bar-On and Moore do, in part because of their fluent use of social knowledge in call interpretation, and in part because of the form of the calls interpreted (structured, discrete, noniconic). Moreover, we are not persuaded of the theoretical utility of the distinction between Gricean and non-Gricean communication, for the reasons discussed in section 1.4. The distinction is formulated within the framework of folk psychology.

5. A grammar overgenerates when it predicts as part of the language sequences that speakers do not take to be part of their language.

6. For example: a male A might visually imagine approaching some female B in the presence of a dominant male C who has a clear line of sight to both A and B. This, in turn, might lead A to predict that the dominant male C will *see* A approaching B, and perhaps that B will *think* A *intends* to mate with that female. In turn, this might lead A to predict that he is likely to be attacked by C if he approaches B, leading to a decision not to do so.

7. For more detail, see Planer (2021).

8. We take no stand on whether increased theory-of-mind capacities in our line reflect innately channeled theory-of-mind mechanisms. Such a view once seemed on firm empirical ground, with only human infants passing implicit false belief tasks. The recent evidence that all great apes possess this capacity (Krupenye, Kano, et al. 2016; Buttelmann, Buttelmann, et al. 2017) complicates that view substantially.

9. As misplaced strikes can cause sharp chips to fly off the core at great speed and in unpredictable directions.

10. As suggested in Tennie, Premo, et al. (2017).

11. The fact that we can do this with a fair degree of accuracy is testimony to how automatized turn-taking becomes in fully competent language users.

12. If one thinks of bee dance as informing the others about their environment, the dances are about the location of distant food. If one thinks of them as imperatives (fly in direction *x* for time *t*), they are about the here and now.

13. There is some controversy over this, for there are claims for a much more creative use of gesture in pantomimes by great apes, including the pantomime by a female chimp to her child of a nut-opening sequence (Russon and Andrews 2011; Russon and Andrews 2015). So there are reports of on-the-fly signal expansion. These reports suggest that when an initial signal failed to elicit the desired response, the signal was not just repeated or even repeated with amplification; it was extended and elaborated; for example, the target of a request is offered a different and better tool with which to comply (for example, a better back-scratching stick). However, these are mostly cases of captive great apes, and they rely on observer interpretations of one-off events, so obviously they need to be treated with much caution. Moreover, some of the reports are at best very marginal cases of pantomime: "Orangutans groomed a partner briefly to solicit grooming; so do chimpanzees and gorillas" (Russon and Andrews 2011, p. 315). So our best guess is that great ape diachronic flexibility is very limited.

14. Identifying the desired response is not at all trivial. If the target of the gesture responds in a way that causes the gesturing agent to stop gesturing, and if that agent does not show obvious signs of frustration and anger, then the response is deemed to have been the intended outcome of the gesture.

15. There is experimental work that shows that iconicity helps greatly when agents have no established medium of communication, but as communication becomes regular, there is a rapid shift to conventional signs (Garrod, Fay, et al. 2007; Fay, Garrod, et al. 2010).

16. The bonobos could produce short and contextually appropriate symbol strings using lexigrams, though their productive capacities were much less extensive than their comprehension. Moreover, since they used a keyboard, it is not clear that they had free recall of the items in their vocabulary. Even so, they could produce as well

as respond. Chaser was a remarkable dog, but without known productive capacity (Pilley 2013).

17. For a very clear recent survey, see the first chapter of Maslin (2017).

18. These tracks are dated to about 5.7 mya. It remains unclear whether the track makers were hominin, but even if they were gorilline, they still illustrate the idea that bipedalism can arise in the great ape stock under a range of different selective regimes (Crompton 2017; Gierliński, Niedźwiedzki, et al. 2017).

19. In the modern case this is probably because they would slow the party down, as ethnographically known foragers have far less to fear from predators than their Pliocene ancestors.

20. Though as one of our readers pointed out, a biped would have roughly double the foot-load per foot compared to a quadruped, presumably imposing a handicap on muddy or sandy substrates.

21. To a skilled reader, they carry a lot of information, not just about the identity of the maker but about its physical and psychological state. An agitated or fearful animal leaves traces unlike those of an animal in calm motion.

22. For a good overview of the challenge of the Pleistocene, see Richerson and Boyd (2013); for a more deep-time perspective dating back to the Cretaceous, see Zachos, Pagani, et al. (2001); for a view on hominin evolution that places central weight on the increasing variability of the Pleistocene, see Potts (1996); Potts (1998); Potts and Faith (2015).

23 Wallace was an externalist, skeptical of sexual selection and perhaps for this reason skeptical of the ability to explain human intelligence through the actions of natural selection. Wallace thought humans had more intelligence than they needed for navigating the external environment successfully (Wallace 1891, chapter 9). Bickerton (2014) followed Wallace in thinking that language poses some special problem for evolutionary theory, as our linguistic resources exceed those we need for subsistence and technical activities. Perhaps they do. But hominins had to survive and flourish in a social world too. In chapter 7, we detail the social factors that selected for elaborated language.

Chapter 3

1. There is recent isotope evidence of a dietary shift toward more meat after the evolution of the erectines, but this does not discriminate between hunting and scavenging as meat sources: see Patterson, Braun, et al. (2019).

2. That said, if Pliocene hominins used naturally occurring cobbles to break open large bones, it is not obvious that this would leave a trace in the historical record. See

Thompson, Carvalho, et al. (2019) for an alternative account that makes scavenging much more foundational to the evolution of the carnivore-omnivore lifeway. We take no stand on this issue.

3. The term "power scavenging" refers to the act of securing access to a carcass by actively driving other predators away from it. Picture a large group of hooting erectines hurling stones at predators feasting on freshly killed ungulate.

4. That said, Karen Lupo (2012) warns us not to rely too heavily on the relationship between scar patterns and tooth patterns in inferences about who got to eat first, as different patterns of access can result in similar patterns on the bones.

5. One example Binford gives is of aboriginal kangaroo hunting, exploiting a quirk in kangaroo predator detection, reducing apparent motion by stalking directly from the front. Another example is that of exploiting horse escape behavior, taking advantage of the fact that horses' default behavior is to run uphill.

6. In response, Pickering and Bunn have argued that endurance hunting is an option only in a narrow range of environments (Pickering and Bunn 2007). We find their case persuasive, though it, in turn, is contested (Lieberman, Bramble, et al. 2007).

7. Gut, like brain, is expensive tissue, and so sacrificing gut can make brain more affordable, at the price of diet change (Aiello and Wheeler 1995).

8. He argues that hominins domesticated fire early, but pounding and soaking are other ways to make USOs more readily edible.

9. This case is made in much more detail in Sterelny (2012a).

10. In discussing hominin life histories, it is important to distinguish the life expectancy of a hominin entering adulthood from life expectancy at birth, as this is often seriously depressed by early childhood mortality.

11. Guthrie (2007, p. 140) estimates the generation length of an elephant at 17 years, of a Pleistocene hominin at 14 (perhaps a year too long), of a white rhino at 7, of a giraffe and a buffalo at 6. Elephants suffer perhaps a 1% adult loss by predation; giraffes 4%; buffalo 5%. Guthrie estimates (p. 141) that the sustainable annual loss rate for Pleistocene hominins could not have been greater than 3%, to make their life histories sustainable. While these figures are obviously estimates, the qualitative point is surely right: slow life histories go with long lifespans, which require very effective means of reducing predation risk.

12. Though not every early *erectus* site evidences the new technology: the North African and the early European erectines seem to miss it (Finlayson 2014).

13. There is some controversy about this. Corbey and colleagues suggest that Acheulian handaxe-making was genetically canalized. Tennie and colleagues accept that some social learning is involved, but of a minimal kind. There is considerable archaeological pushback against these ideas: for example in the commentaries on

Tennie's paper, and Shipton's demonstration of social learning in the Acheulian. See Corbey, Jagich, et al. (2016); Tennie, Premo, et al. (2017); Shipton (2019).

14. Or aspects of the action being recommended, for imperative signals.

15. However, inference about vocal control from fossil evidence about the shape of the larynx, the size of the spinal canal, and the size of hypoglossal canals through which the nerves controlling tongue movement pass is difficult and error-prone. (More on this in chapter 6.) For a brief, clear, and very measured discussion see Hurford (2014).

16. We do not claim that earlier hominins and other animals never represent future options and needs; we just claim that this was a much more central feature of erectine cognition.

17. See the contrast between, e.g., Devitt (1981), Devitt (1996) and Jackson (1997), Jackson (2010).

18. This may well include learning the meaning of words. The "natural pedagogy" group have a quite detailed picture of how this works with young children (Csibra and Gergely 2009; Csibra and Gergely 2011).

19. The great genetic similarity of AMHs, Neanderthals, and Denisovans suggests that all three were also similar to their common ancestor, *heidelbergensis*. See Dediu and Levinson (2013); Levinson and Dediu (2018), who place great weight on this inference in their argument for the early evolution of language.

Chapter 4

1. Rarely unless the time, place, and sender are taken to be part of the signal itself. We think not, in the typical case: see section 1.4. Unless the sender deliberately places himself/herself at a particular place and time, location is a cue, not part of the signal. Consider the contrast between a typical signal sent in response to some unanticipated encounter and a political leader speaking from an elevated position above a group. In the second case, location is an aspect of the signal. In addition, as one of our readers pointed out, whether structured signals are common depends on what counts as a single sign. When a dog invites play by bowing and play-barking, is that a composite sign or a single physically complex sign? We incline to the second view, because elements of the display are not used independently of one another. See also Scott-Phillips and Blythe (2013), who use the term "combinatorial signal" for what we are here calling "composite signals."

2. There is evidence of a fair degree of developmental plasticity with apes' pointing repertoires: see Leavens, Hopkins, et al. (2005).

3. Note that, even here, we are not attributing to the sender an ability to think recursively about the receiver's mental states, i.e., to think about the receiver's thinking about the sender's own thinking.

4. Goldin-Meadow notes that children acquiring a conventional spoken language typically pass through a similar stage, even when the language they are acquiring does not feature a lexical distinction between nouns, verbs, and adjectives (not all languages clearly do).

5. Goldin-Meadow and colleagues show how, in an experimental setting, the consistency in gestural forms decreases over time as a result of poor comprehension on the part of communicative partners.

Chapter 5

1. The St. Andrews team's methodology assigns meaning based on so-called "apparently satisfactory outcomes" or ASOs. In principle, a sender might produce a series where the intended meaning was some function of the parts, but the receiver might simply attend to and comply with the meaning of the first or last sign in the series, for example. If senders settle for this outcome, as better than nothing, then we would miss the richness inherent in the original sequence.

2. In the neuroscience jargon, "caudal" means toward the posterior part of the body, whereas "rostral" means toward the front part of the body. Due to humans' upright posture, these terms effectively mean toward the front and back of the brain.

3. Some of these neurons, so-called "canonical F5 neurons," respond to specific action *affordances*, e.g., objects that permit grabbing, while others are of the more standard mirror neuron type.

4. Apparently, artifacts without these features would have served utilitarian purposes equally well (or even better, by comparison to the awkwardly large tools). So this change in artifact form is seen as evidence that these tools had become social signals, as they could have been made only by the highly skilled and knowledgeable. The manufacture of such tools served as a hard-to-fake signal of expertise (Hiscock 2014). In one version of the idea, it becomes a signal of mate quality (as in Kohn and Mithen's [1999] "sexy handaxe theory").

5. The term "platform" refers to a specific part of a core from which a flake is to be struck (or from which a flake has been struck).

6. A simple example of such a convention can be illustrated with the sentence "Intuitively, birds that fly swim" (an example of Chomsky's). English speakers interpret "intuitively" as modifying "swim," not "fly," despite the fact that "intuitively" is closer to "fly" in terms of linear distance. To explain this pattern, it is (arguably) necessary to understand English speakers as representing the sentence hierarchically. Specifically, the thought is that they represent "intuitively" as separated by less hierarchical depth from "swim" than from "fly." This way of using hierarchical structure in interpretation would be an instance of a hierarchical syntactic convention.

7. See again Frank, Bod, and Christiansen (2012), for example.

Chapter 6

1. It is not hard to imagine how such audience sensitivity to gesture could be very useful in pedagogical contexts.

2. Some theorists (e.g., McNeill 2008) take this point much farther, treating speech and gesture as inseparable components of a larger, integrated communication system.

3. Gibbons are well known for their impressive singing, but there is no indication that they acquire songs or parts of songs via imitation.

4. As Wrangham nicely explains, "raw foodists" are in general quite unhealthy. One striking illustration of this is that many female raw foodists stop menstruating.

5. Pounding already cooked food made little difference in the study.

6. https://www.nsc.org/home-safety/safety-topics/choking-suffocation

7. For a nice overview, see Fitch (2010). Fitch is one of the most prominent defenders of this view. For a more nuanced view of the music-language connection, see Killin (2017a); Killin (2017b).

8. The term "firelight niche" is adapted from a phrase used by anthropologist Polly Wiessner in a 2014 article analyzing the fireside conversations of the Ju/'hoan (!Kung) Bushmen of South Africa (she uses the term "firelit niche").

9. There is also a protein change in this gene, but it is unclear whether this change is functionally significant.

Chapter 7

1. A term of art denoting predominantly grass and woodland environments during the Pliocene and Pleistocene which reached from Africa to China.

2. We will return to this in section 7.3, but some attempts at dating and explaining are given in Chapais (2008); Layton and O'Hara (2010); Layton, O'Hara, and Bilsborough (2012); Chapais (2013); Sterelny (2019b).

3. The effectiveness of the latter as a weapons technology should not be underestimated. See Isaac (1987) for a nice discussion, including a discussion of the role of this technology in ethnographically known peoples.

4. Bettinger argues that spear-throwers were team weapons, and that small-group and solitary hunting depended on the bow and arrow, but reports from Australia suggest that Australian Aborigines hunted kangaroo and emus with spear-throwers: Meggitt (1965); Bettinger (2013).

5. See Pinker (2011), though his figures may well exaggerate the risk, for many come from cultures severely dislocated by colonial violence.

6. For a discussion and overview of the relevant experiments, see Tomasello (2016).

7. Gould (1969) published photos of closed rituals, thus violating those norms, and as a result deeply offended local sensibilities and was never able to work in the western desert again (Griffiths 2018).

8. See Wrangham (2019) for a nice overview.

9. Depending on how the details are spelled out, this might not count as a self-domestication theory of human origins. Unless human interactions drive selection, the theory is really a version of a *natural* domestication theory.

10. Compare, for example, Li and Durbin (2011) and Schiffels and Durbin (2014).

11. As always, there are complications. Semiarid plants invest more of their smaller biomass growth in storage and reproductive organs, and these are more valuable to humans as resources. Moreover, cold dry periods tend to favor grasslands over forest. Much of the biomass in forests is unavailable to grazers and even browsers, while, for example, the cold steppelands had a high animal biomass. The inference from climate to population size is fraught (Guthrie 2001).

Chapter 8

1. With apologies to Russell Gray, who also used this title in developing an incremental view of the evolution of language.

References

Aiello, L. C., and P. Wheeler (1995). "The Expensive-Tissue Hypothesis: The Brain and the Digestive System in Human and Primate Evolution." *Current Anthropology* 36 (2).

Alibali, M. W., L. M. Flevares, and S. Goldin-Meadow (1997). "Assessing Knowledge Conveyed in Gesture: Do Teachers Have the Upper Hand?" *Journal of Educational Psychology* 89 (1): 183.

Alperson-Afil, N. (2008). "Continual Fire-Making by Hominins at Gesher Benot Ya'aqov, Israel." *Quaternary Science Reviews* 27 (September): 1733–1739.

Alperson-Afil, N., and N. Goren-Inbar (2010). *The Acheulian Site of Gesher Benot Ya'aqov*. Vol. II: *Ancient Flames and Controlled Use of Fire*. New York: Springer.

Alperson-Afil, N., D. Richter, et al. (2007). "Phantom Hearths and the Use of Fire at Gesher Benot Ya'Aqov, Israel (2007)." *PaleoAnthropology* 3: 1–15.

Alvard, M., and D. Nolin (2002). "Rousseau's Whale Hunt? Coordination among Big Game Hunters." *Current Anthropology* 43 (4): 533–559.

Ambrose, S. (2001). "Paleolithic Technology and Human Evolution." *Science* 291 (March 2): 1748–1753.

Ambrose, S. H. (2010). "Coevolution of Composite-Tool Technology, Constructive Memory, and Language: Implications for the Evolution of Modern Human Behavior." *Current Anthropology* 51 (S1): S135–S147.

Arbib, M. A., K. Liebal, and S. Pika (2008). "Primate Vocalization, Gesture, and the Evolution of Human Language." *Current Anthropology* 49 (6): 1053–1076.

Austin, J. (1962). *How to Do Things with Words*. Oxford: Oxford University Press.

Baars, B. J. (1997). "In the Theatre of Consciousness: Global Workspace Theory, a Rigorous Scientific Theory of Consciousness." *Journal of Consciousness Studies* 4 (4): 292–309.

Baars, B. J. (2005). "Global Workspace Theory of Consciousness: Toward a Cognitive Neuroscience of Human Experience." *Progress in Brain Research* 150: 45–53.

Badre, D., and M. D'Esposito (2009). "Is the Rostro-caudal Axis of the Frontal Lobe Hierarchical?" *Nature Reviews Neuroscience* 10 (9): 659.

Badre, D., J. Hoffman, et al. (2009). "Hierarchical Cognitive Control Deficits Following Damage to the Human Frontal Lobe." *Nature Neuroscience* 12 (4): 515.

Barkai, R., J. Rosell, et al. (2017). "Fire for a Reason: Barbecue at Middle Pleistocene Qesem Cave, Israel." *Current Anthropology* 58 (S16).

Bar-On, D., and R. Moore (2017). "Pragmatic Interpretation and Signaler-Receiver Asymmetries in Animal Communication." In *The Routledge Handbook of Philosophy of Animal Minds*, ed. K. Andrews and J. Beck, 291–300. New York: Routledge.

Bates, E., and B. McWhinney (1982). "Functionalist Approaches to Grammar." In *Language Acquisition: The State of the Art*, ed. E. Wanne and L. Gleitman, 173–218. Cambridge: Cambridge University Press.

Beckerman, S., and P. Valentine, eds. (2002). *Cultures of Multiple Fathers: The Theory and Practice of Partible Paternity in Lowland South America*. Gainesville: University Press of Florida.

Belyaev, D. K., A. O. Ruvinsky, and L. N. Trut (1981). "Inherited Activation-Inactivation of the Star Gene in Foxes: Its Bearing on the Problem of Domestication." *Journal of Heredity* 72 (4): 267–274.

Ben-Dor, M., A. Gopher, et al. (2011). "Man the Fat Hunter: The Demise of *Homo erectus* and the Emergence of a New Hominin Lineage in the Middle Pleistocene (ca. 400 kyr) Levant." *PLoS One* 6 (12): e28689.

Berwick, R. C., and N. Chomsky (2016). *Why Only Us: Language and Its Evolution*. Cambridge, MA: MIT Press.

Bettinger, R. (2013). "Effects of the Bow on Social Organization in Western North America." *Evolutionary Anthropology: Issues, News, and Reviews* 22 (3): 118–123.

Bickerton, D. (2002). "From Protolanguage to Language." In *The Speciation of Modern Homo sapiens*, ed. T. J. Crow, 193–120. Oxford: Oxford University Press.

Bickerton, D. (2009). *Adam's Tongue: How Humans Made Language, How Language Made Humans*. New York: Hill and Wang.

Bickerton, D. (2014). *More Than Nature Needs: Language, Mind and Evolution*. Cambridge, MA: Harvard University Press.

Binford, L. (1980). "Willow Smoke and Dogs' Tails: Hunter-Gatherer Settlement Systems and Archaeological Site Formation." *American Antiquity* 45 (1): 4–20.

Binford, L. (2007). "The Diet of Early Hominins: Some Things We Need to Know Before 'Reading' the Menu from the Archaeological Record." In *Guts and Brains*, ed. W. Roebroeks, 185–222. Leiden: Leiden University Press.

Bingham, P. (1999). "Human Uniqueness: A General Theory." *Quarterly Review of Biology* 74 (2): 133–169.

Bingham, P. (2000). "Human Evolution and Human History: A Complete Theory." *Evolutionary Anthropology* 9 (6): 248–257.

Boeckx, C. A., and K. Fujita (2014). "Syntax, Action, Comparative Cognitive Science, and Darwinian Thinking." *Frontiers in Psychology* 5: 627.

Boehm, C. (1999). *Hierarchy in the Forest.* Cambridge, MA: Harvard University Press.

Boehm, C. (2012). *Moral Origins: The Evolution of Virtue, Altruism, and Shame.* New York: Basic Books.

Boesch, C., and H. Boesch (1983). "Optimisation of Nut-Cracking with Natural Hammers by Wild Chimpanzees." *Behaviour* 83 (3–4): 265–286.

Boesch, C., J. Head, and M. M. Robbins (2009). "Complex Tool Sets for Honey Extraction among Chimpanzees in Loango National Park, Gabon." *Journal of Human Evolution* 56 (6): 560–569.

Bohn, M., J. Call, et al. (2015). "Communication about Absent Entities in Great Apes and Human Infants." *Cognition* 145: 63–72.

Bohn, M., J. Call, et al. (2016). "The Role of Past Interactions in Great Apes' Communication about Absent Entities." *Journal of Comparative Psychology* 130 (4): 351–357.

Bonta, M., R. Gosford, et al. (2017). "Intentional Fire-Spreading by 'Firehawk' Raptors in Northern Australia." *Journal of Ethnobiology* 37 (4): 700–718.

Botvinick, M. M. (2008). "Hierarchical Models of Behavior and Prefrontal Function." *Trends in Cognitive Sciences* 12 (5): 201–208.

Boyd, R. (2016). *A Different Kind of Animal: How Culture Made Humans Exceptionally Adaptable and Cooperative.* Princeton: Princeton University Press.

Brainard, G. C., J. P. Hanifin, et al. (2001). "Action Spectrum for Melatonin Regulation in Humans: Evidence for a Novel Circadian Photoreceptor." *Journal of Neuroscience* 21 (16): 6405–6412.

Bruner, E. (2010). "Morphological Differences in the Parietal Lobes within the Human Genus: A Neurofunctional Perspective." *Current Anthropology* 51 (S1): S77–S88.

Bullinger, A. F., F. Zimmerman, et al. (2011). "Different Social Motives in the Gestural Communication of Chimpanzees and Human Children." *Developmental Science* 14 (1): 58–68.

Bunn, H. (2007). "Meat Made Us Human." In *Evolution of the Human Diet: The Known, the Unknown, and the Unknowable*, ed. P. Ungar, 191–211. Oxford: Oxford University Press.

Bunn, H., and A. Gurtov (2014). "Prey Mortality Profiles Indicate that Early Pleistocene Homo at Olduvai Was an Ambush Predator." *Quaternary International* 322: 44–53.

Bunn, H., and T. R. Pickering (2010). "Bovid Mortality Profiles in Paleoecological Context Falsify Hypotheses of Endurance Running–Hunting and Passive Scavenging by Early Pleistocene Hominins." *Quaternary Research* 74 (3): 395–404.

Bureau of Labor Statistics (2020). "American Time Use Survey—Results 2019." US Department of Labor. https://www.bls.gov/news.release/pdf/atus.pdf.

Buttelmann, D., F. Buttelmann, et al. (2017). "Great Apes Distinguish True from False Beliefs in an Interactive Helping Task." *PLoS One* 12 (4).

Byrne, R. W., E. Cartmill, et al. (2017). "Great Ape Gestures: Intentional Communication with a Rich Set of Innate Signals." *Animal Cognition* 20 (4): 755–769.

Calcott, B. (2008). "Lineage Explanations: Explaining How Biological Mechanisms Change." *British Journal for the Philosophy of Science* 60: 51–78.

Capasso, L., E. Michetti, and R. D'Anastasio (2008). "A *Homo erectus* Hyoid Bone: Possible Implications for the Origin of the Human Capability for Speech." *Collegium Antropologicum* 32 (4): 1007–1011.

Carmody, R. N., G. S. Weintraub, and R. W. Wrangham (2011). "Energetic Consequences of Thermal and Nonthermal Food Processing." *Proceedings of the National Academy of Sciences* 108 (48): 19199–19203.

Carrigan, E. M., and A. M. Coppola (2017). "Successful Communication Does Not Drive Language Development: Evidence from Adult Homesign." *Cognition* 158: 10–27.

Carruthers, P. (2015). *The Centered Mind: What the Science of Working Memory Shows Us about the Nature of Human Thought.* Oxford: Oxford University Press.

Carvalho, S., E. Cunha, et al. (2008). "Chaînes Opératoires and Resource-Exploitation Strategies in Chimpanzee (*Pan troglodytes*) Nut Cracking." *Journal of Human Evolution* 55 (1): 148–163.

Cashdan, E. (1980). "Egalitarianism among Hunters and Gatherers." *American Anthropologist* 82: 116–120.

Chapais, B. (2008). *Primeval Kinship*. Cambridge, MA: Harvard University Press.

Chapais, B. (2013). "Monogamy, Strongly Bonded Groups and the Evolution of Human Social Structure." *Evolutionary Anthropology* 22: 52–65.

Christensen, W., and J. Michael (2015). "From Two Systems to a Multi-systems Architecture for Mindreading." *New Ideas in Psychology* 40: 48–64.

Christiansen, M., and N. Chater (2016a). *Creating Language: Integrating Evolution, Acquisition, and Processing*. Cambridge, MA: MIT Press 2016.

Christiansen, M. H., and N. Chater (2016b). "The Now-or-Never bottleneck: A Fundamental Constraint on Language." *Behavioral and Brain Sciences* 39.

Christoff, K., K. Keramatian, et al. (2009). "Prefrontal Organization of Cognitive Control According to Levels of Abstraction." *Brain Research* 1286: 94–105.

Churchill, S., and J. Rhodes (2009). "The Evolution of the Human Capacity for 'Killing at a Distance': The Human Fossil Evidence for the Evolution of Projectile Weaponry." In *The Evolution of Hominin Diets: Integrating Approaches to the Study of Palaeolithic Subsistence*, ed. J.-J. Hublin and M. Richards, 201–210. Dordrecht: Springer.

Cieri, R. L., S. E. Churchill, et al. (2014). "Craniofacial Feminization, Social Tolerance, and the Origins of Behavioral Modernity." *Current Anthropology* 55 (4): 419–443.

Claidiere, N., K. Smith, et al. (2014). "Cultural Evolution of Systematically Structured Behaviour in a Non-human Primate." *Proceedings of the Royal Society* 281.

Coolidge, F. L., and T. Wynn (2005). "Working Memory, Its Executive Functions, and the Emergence of Modern Thinking." *Cambridge Archaeological Journal* 15 (1): 5–26.

Coolidge, F. L., and T. G. Wynn (2018). *The Rise of Homo sapiens: The Evolution of Modern Thinking*. Oxford: Oxford University Press.

Corballis, M. (2009). "The Evolution of Language." *Annals of the New York Academy of Sciences* 1156 (March): 19–43.

Corballis, M. C. (2011). *The Recursive Mind: The Origins of Human Language, Thought, and Civilization*. Princeton: Princeton University Press.

Corballis, M. C. (2014). *The Recursive Mind: The Origins of Human Language, Thought, and Civilization*. Updated ed. Princeton: Princeton University Press.

Corbey, R., A. Jagich, et al. (2016). "The Acheulean Handaxe: More Like a Bird's Song Than a Beatles' Tune?" *Evolutionary Archaeology* 25: 6–19.

Cordell, J. (1974). "The Lunar-Tide Fishing Cycle in Northeastern Brazil." *Ethnology* 13 (4): 379–392.

Crockford, C. (2019). "Why Does the Chimpanzee Vocal Repertoire Remain Poorly Understood? And What Can Be Done about It." In *The Tai Chimpanzees: 40 Years of Research*, ed. C. Boesch and R. Wittig. Cambridge: Cambridge University Press.

Crockford, C., R. M. Wittig, et al. (2012). "Wild Chimpanzees Inform Ignorant Group Members of Danger." *Current Biology* 22 (January 24): 142–146.

Crockford, C., R. Wittig, et al. (2017). "Vocalizing in Chimpanzees Is Influenced by Social-Cognitive Processes." *Science Advances* 3 (11).

Crompton, R. H. (2017). “Making the Case for Possible Hominin Footprints from the Late Miocene (c. 5.7 Ma) of Crete?” *Proceedings of the Geologists' Association* 128 (5–6): 692–693.

Csibra, G., and G. Gergely (2009). “Natural Pedagogy.” *Trends in Cognitive Science* 13 (4): 148–153.

Csibra, G., and G. Gergely (2011). “Natural Pedagogy as Evolutionary Adaptation.” *Philosophical Transactions of the Royal Society B* 366 (1567): 1149–1157.

Deacon, T. (1997). *The Symbolic Species: The Co-evolution of Language and the Brain*. New York: W. W. Norton.

de Boer, B. (2009). “Acoustic Analysis of Primate Air Sacs and Their Effect on Vocalization.” *Journal of the Acoustical Society of America* 126 (6): 3329–3343.

Dediu, D., and S. Levinson (2013). “On the Antiquity of Language: The Reinterpretation of Neandertal Linguistic Capacities and Its Consequences.” *Frontiers of Psychology* 4 (397).

Dennett, D. C. (1975). “Three Kinds of Intentional Psychology.” In *Reduction, Time, and Reality: Studies in the Philosophy of the Natural Sciences*, ed. R. Healy. Cambridge: Cambridge University Press.

Dennett, D. C. (1991). “Real Patterns.” *Journal of Philosophy* 87: 27–51.

Dennett, D. (2017). *From Bacteria to Bach and Back*. New York: W. W. Norton.

Dessalles, J.-L. (2007). *Why We Talk: The Evolutionary Origins of Language*. Oxford: Oxford University Press.

Devitt, M. (1981). *Designation*. New York: Columbia University Press.

Devitt, M. (1996). *Coming to Our Senses: A Naturalistic Program for Semantic Localism*. Cambridge: Cambridge University Press.

Dezecache, G., and R. I. Dunbar (2012). “Sharing the Joke: The Size of Natural Laughter Groups.” *Evolution and Human Behavior* 33 (6): 775–779.

Dixon, M. L., K. C. Fox, and K. Christoff (2014). “Evidence for Rostro-caudal Functional Organization in Multiple Brain Areas Related to Goal-Directed Behavior.” *Brain Research* 1572: 26–39.

Domínguez-Rodrigo, M., and T. R. Pickering (2017). “The Meat of the Matter: An Evolutionary Perspective on Human Carnivory.” *Azania: Archaeological Research in Africa* 52 (1): 4–32.

Donald, M. (1991). *Origins of the Modern Mind: Three Stages in the Evolution of Culture and Cognition*. Cambridge, MA: Harvard University Press.

Donald, M. (2001). *A Mind So Rare: The Evolution of Human Consciousness*. New York: W. W. Norton.

Dunbar, R. (1996). *Grooming, Gossip and the Evolution of Language*. London: Faber and Faber.

Dunbar, R. (2009). "Why Only Humans Have Language." In *The Prehistory of Language*, ed. R. Botha and C. Knight, 12–35. Oxford: Oxford University Press.

Dunbar, R. (2014). "Why Only Humans Have Language." In *Lucy to Language: The Benchmark Papers*, ed. R. Dunbar, C. Gamble, and J. Gowlett, 426–445. Oxford: Oxford University Press.

Dunbar, R. I. M. (2017). "Group Size, Vocal Grooming and the Origins of Language." *Psychonomic Bulletin and Review* 24 (1): 209–212.

Dunbar, R. I., R. Baron, et al. (2012). "Social Laughter Is Correlated with an Elevated Pain Threshold." *Proceedings of the Royal Society B: Biological Sciences* 279 (1731): 1161–1167.

Dunbar, R., and J. Gowlett (2014). "Fireside Chat: The Impact of Fire on Hominin Socioecology." In *Lucy to Language: The Benchmark Papers*, ed. R. Dunbar, C. Gamble, and J. Gowlett, 277–296. Oxford: Oxford University Press.

Evans, N. (2017). "Did Language Evolve in Multilingual Settings?" *Biology and Philosophy* 32 (5): 905–933.

Evans-Pritchard, E. E. (1940). *The Nuer: A Description of the Modes of Livelihood and Political Institutions of a Nilotic People*. Oxford: Clarendon Press.

Everett, D. L. (2005). "Cultural Constraints on Grammar and Cognition in Pirahã: Another Look at the Design Features of Human Language." *Current Anthropology* 46 (4): 621–646.

Everett, D. (2009). "Pirahã Culture and Grammar: A Response to Some Criticisms." *Language* 85 (2): 405–442.

Everett, D. (2017). *How Language Began*. London: Profile Books.

Fadiga, L., L. Craighero, and A. D'Ausilio (2009). "Broca's Area in Language, Action, and Music." *Annals of the New York Academy of Sciences* 1169 (1): 448–458.

Fay, N., M. Ellison, et al. (2015). "Iconicity: From Sign to System in Human Communication and Language." *Pragmatics and Cognition* 22 (2): 244–263.

Fay, N., S. Garrod, et al. (2010). "The Interactive Evolution of Human Communication Systems." *Cognitive Science* 34 (3): 351–386.

Ferrari, P. F., V. Gallese, et al. (2003). "Mirror Neurons Responding to the Observation of Ingestive and Communicative Mouth Actions in the Monkey Ventral Premotor Cortex." *European Journal of Neuroscience* 17 (8): 1703–1714.

Finlayson, C. (2014). *The Improbable Primate: How Water Shaped Human Evolution*. New York: Oxford University Press.

Fitch, W. T. (2000). "The Evolution of Speech: A Comparative Review." *Trends in Cognitive Sciences* 4 (7): 258–267.

Fitch, W. T. (2010). *The Evolution of Language.* Cambridge: Cambridge University Press.

Flynn, J. P. (1967). "The Neural Basis of Aggression in Cats." In *Neurophysiology and Emotion*, ed. David C. Glass, 40–60. New York: Rockefeller University Press.

Fodor, J. A. (1975). *The Language of Thought.* New York: Thomas Y. Crowell.

Fodor, J. A. (1983). *The Modularity of Mind.* Cambridge, MA: MIT Press.

Fonseca-Azevedo, K., and S. Herculano-Houzel (2012). "Metabolic Constraint Imposes Tradeoff between Body Size and Number of Brain Neurons in Human Evolution." *Proceedings of the National Academy of Sciences* 109 (45): 18571–18576.

Frank, S. L., R. Bod, and M. H. Christiansen (2012). "How Hierarchical Is Language Use?" *Proceedings of the Royal Society B: Biological Sciences* 279 (1747): 4522–4531.

Frison, G. C. (2004). *Survival by Hunting: Prehistoric Human Predators and Animal Prey.* Berkeley: University of California Press.

Fröhlich, M., R. M. Wittig, and S. Pika (2016). "Should I Stay or Should I Go? Initiation of Joint Travel in Mother–Infant Dyads of Two Chimpanzee Communities in the Wild." *Animal Cognition* 19 (3): 483–500.

Furuichi, T. (2011). "Female Contributions to the Peaceful Nature of Bonobo Society." *Evolutionary Anthropology: Issues, News, and Reviews* 20 (4): 131–142.

Galdikas, B. M. (1988). "Orangutan Diet, Range, and Activity at Tanjung Puting, Central Borneo." *International Journal of Primatology* 9 (1): 1–35.

Gallese, V., L. Fadiga, et al. (1996). "Action Recognition in the Premotor Cortex." *Brain* 119 (2): 593–609.

Gamble, C. (1998). "Palaeolithic Society and the Release from Proximity: A Network Approach to Intimate Relations." *World Archaeology* 29: 426–449.

Gamble, C. (2008). "Kinship and Material Culture: Archaeological Implications of the Human Global Diaspora." In *Early Human Kinship: From Sex to Social Reproduction*, ed. N. Allen et al., 27–40. Oxford: Blackwell.

Gamble, C. (2013). *Settling the Earth: The Archaeology of Deep Human History.* Cambridge: Cambridge University Press.

Gamble, C., J. Gowlett, and R. Dunbar (2014). *Thinking Big: How the Evolution of Social Life Shaped the Human Mind.* London: Thames and Hudson.

Garde, M. (2009). "The Language of Fire: Seasonality, Resources and Landscape Burning on the Arnhem Land Plateau." In *Culture, Ecology and Economy of Fire*

Management in North Australian Savannas, ed. J. Russell-Smith, P. Whitehead, and P. Cooke, 85–164. Collingwood: CSIRO Publishing.

Garrod, S., N. Fay, et al. (2007). "Foundations of Representation: Where Might Graphical Symbol Systems Come From?" *Cognitive Science* 31 (6): 961–987.

Genty, E., T. Breuer, et al. (2009). "Gestural Communication of the Gorilla (*Gorilla gorilla*): Repertoire, Intentionality and Possible Origins." *Animal Cognition* 12 (3): 527–546.

Genty, E., and R. W. Byrne (2010). "Why Do Gorillas Make Sequences of Gestures?" *Animal Cognition* 13 (2): 287–301.

Gergely, G., and G. Csibra (2006). "Sylvia's Recipe: The Role of Imitation and Pedagogy in the Transmission of Cultural Knowledge." In *Roots of Human Society: Culture, Cognition and Human Interaction*, ed. N. J. Enfield and S. C. Levenson, 229–255. Oxford: Berg.

Gergely, G., K. Egyed, et al. (2007). "On Pedagogy." *Developmental Science* 10 (1): 139–146.

Ghiglieri, M. P. (1984). *The Chimpanzees of Kibale Forest: A Field Study of Ecology and Social Structure*. New York: Columbia University Press.

Gierliński, G. D., G. Niedźwiedzki, et al. (2017). "Possible Hominin Footprints from the Late Miocene (c. 5.7 Ma) of Crete?" *Proceedings of the Geologists' Association* 128 (5–6): 697–710.

Gleeson, B. T., and G. Kushnick (2018). "Female Status, Food Security, and Stature Sexual Dimorphism: Testing Mate Choice as a Mechanism in Human Self-Domestication." *American Journal of Physical Anthropology* 167 (3): 458–469.

Godfrey-Smith, P. (1996). *Complexity and the Function of Mind in Nature*. Cambridge: Cambridge University Press.

Godfrey-Smith, P. (2017). "Senders, Receivers, and Symbolic Artifacts." *Biological Theory* 12 (4): 275–286.

Gokhman, D., M. Nissim-Rafinia, et al. (2019). "Genes Affecting Vocal and Facial Anatomy Went through Extensive Regulatory Divergence in Modern Humans." *bioRxiv*, 106955.

Goldin-Meadow, S. (1999). "The Role of Gesture in Communication and Thinking." *Trends in Cognitive Sciences* 3 (11): 419–429.

Goldin-Meadow, S. (2002). "Getting a Handle on Language Creation." *Typological Studies in Language* 53: 343–374.

Goldin-Meadow, S. (2003). *The Resilience of Language. Essays in Developmental Psychology*. New York: Psychology Press.

Goldin-Meadow, S. (2015). "Watching Language Grow in the Manual Modality: Nominals, Predicates, and Handshapes." *Cognition* 136: 381–395.

Goldin-Meadow, S., and C. Mylander (1984). "Gestural Communication in Deaf Children and the Effects and Noneffects of Parental Input." *Monographs of the Society for Research in Child Development* 49 (3–4): 1–151.

Goldin-Meadow, S., and C. M. Sandhofer (1999). "Gestures Convey Substantive Information about a Child's Thoughts to Ordinary Listeners." *Developmental Science* 2 (1): 67–74.

Goldin-Meadow, S., D. Wein, and C. Chang (1992). "Assessing Knowledge through Gesture: Using Children's Hands to Read Their Minds." *Cognition and Instruction* 9 (3): 201–219.

Goren-Inbar, N., N. Alperson, et al. (2004). "Evidence of Hominin Control of Fire at Gesher Benot Yaaqov, Israel." *Science* 304 (5671): 725–727.

Gough, E. K. (1959). "The Nayars and the Definition of Marriage." *Journal of the Royal Anthropological Institute of Great Britain and Ireland* 89 (1): 23–34.

Gould, R. A. (1969). *Yiwara: Foragers of the Australian Desert.* London: Collins.

Gowlett, J. A. J. (2006). "The Early Settlement of Northern Europe: Fire History in the Context of Climate Change and the Social Brain." *Comptes Rendus Palevol* 5: 299–310.

Gowlett, J. (2016). "The Discovery of Fire by Humans: A Long and Convoluted Process." *Philosophical Transactions of the Royal Society B* 371 (1696).

Gowlett, J., and R. Wrangham (2013). "Earliest Fire in Africa: Towards the Convergence of Archaeological Evidence and the Cooking Hypothesis." *Azania: Archaeological Research in Africa* 48 (1): 5–30.

Green, E. M. (2017). "Performing Gesture: The Pragmatic Functions of Pantomimic and Lexical Repertoires in a Natural Sign Narrative." *Gesture* 16 (2): 329–363.

Green, J. (2016). *Drawn from the Ground: Sound, Sign and Inscription in Central Australian Sand Stories.* Cambridge: Cambridge University Press.

Griffiths, B. (2018). *Deep Time Dreaming: Uncovering Ancient Australia.* Melbourne: Black Ink.

Guthrie, R. D. (2001). "Origin and Causes of the Mammoth Steppe: A Story of Cloud Cover, Woolly Mammoth Tooth Pits, Buckles, and Inside-Out Beringia." *Quaternary Science Reviews* 20: 549–574.

Guthrie, R. D. (2007). "Haak en Steek—The Tool that Allowed Hominins to Colonize the African Savanna and Flourish There." In *Guts and Brains*, ed. W. Roebroeks, 133–164. Leiden: Leiden University Press.

Hale, K. (1983). "Warlpiri and the Grammar of Non-configurational Languages." *Natural Language and Linguistic Theory* 1 (1): 5–47.

Harmand, S., J. E. Lewis, et al. (2015). "3.3-Million-Year-Old Stone Tools from Lomekwi 3, West Turkana, Kenya." *Nature* 521 (7552): 310–315.

Hart, C., and H. Pilling (1960). *The Tiwi of North Australia*. New York: Holt Rinehart Winston.

Hart, D., and R. W. Sussman (2005). *Man the Hunted: Primates, Predators, and Human Evolution*. New York: Westview Press.

Hauser, M. (1996). *The Evolution of Communication*. Cambridge, MA: MIT Press.

Hauser, M., N. Chomsky, et al. (2002). "The Faculty of Language: What Is It, Who Has It, and How Does It Evolve?" *Science* 298: 1569–1579.

Hawkes, K. (1994). "The Grandmother Effect." *Nature* 428 (11 March): 128–129.

Hawkes, K. (2003). "Grandmothers and the Evolution of Human Longevity." *American Journal of Human Biology* 15 (3): 380—400.

Hawkes, K., and R. Bird (2002). "Showing Off, Handicap Signaling and the Evolution of Men's Work." *Evolutionary Anthropology* 11 (1): 58–67.

Hawkes, K., J. F. O'Connell, et al. (1998). "Grandmothering, Menopause and the Evolution of Human Life Histories." *Proceedings of the National Academy of Sciences* 95: 1336–1339.

Hawkes, K., J. F. O'Connell, et al. (2010). "Family Provisioning Is Not the Only Reason Men Hunt." *Current Anthropology* 51 (2): 259–264.

Henrich, J. (2004). "Demography and Cultural Evolution: Why Adaptive Cultural Processes Produced Maladaptive Losses in Tasmania." *American Antiquity* 69 (2): 197–221.

Henrich, J. (2016). *The Secret of Our Success: How Culture Is Driving Human Evolution, Domesticating Our Species and Making Us Smarter*. Princeton: Princeton University Press.

Henshilwood, C., and C. Marean (2003). "The Origin of Modern Human Behavior." *Current Anthropology* 44: 627–651.

Herculano-Houzel, S. (2016). *The Human Advantage: A New Understanding of How Our Brain Became Remarkable*. Cambridge, MA: MIT Press.

Heyes, C. (2011). "Automatic Imitation." *Psychological Bulletin* 137 (3): 463–483.

Heyes, C. (2012). "Grist and Mills: On the Cultural Origins of Cultural Learning." *Philosophical Transactions of the Royal Society B* 367: 2181–2191.

Heyes, C. (2018). *Cognitive Gadgets: The Cultural Evolution of Thinking*. Cambridge, MA: Harvard University Press.

Heyes, C., and C. Frith (2014). "The Cultural Evolution of Mind Reading." *Science* 344 (6190).

Hill, K., M. Barton, et al. (2009). "The Emergence of Human Uniqueness: Characters Underlying Behavioral Modernity." *Evolutionary Anthropology* 18: 174–187.

Hill, K., R. S. Walker, et al. (2011). "Co-residence Patterns in Hunter-Gatherer Societies Show Unique Human Social Structure." *Science* 331 (6022): 1286.

Hiscock, P. (2014). "Learning in Lithic Landscapes: A Reconsideration of the Hominid 'Toolusing' Niche." *Biological Theory* 9 (1): 27–41.

Hobaiter, C., and R. Byrne (2014). "The Meanings of Chimpanzee Gestures." *Current Biology* 24 (July 21): 1596–1600.

Hostetter, A. B., and M. W. Alibali (2008). "Visible Embodiment: Gestures as Simulated Action." *Psychonomic Bulletin and Review* 15 (3): 495–514.

Hrdy, S. B. (2009). *Mothers and Others: The Evolutionary Origins of Mutual Understanding.* Cambridge, MA: Harvard University Press.

Hublin, J. J., A. Ben-Ncer, et al. (2017). "New Fossils from Jebel Irhoud, Morocco and the Pan-African Origin of *Homo sapiens*." *Nature* 546 (7657): 289–292.

Hurford, J. (2007). *The Origins of Meaning.* New York: Oxford University Press.

Hurford, J. (2011). *The Origins of Grammar: Language in the Light of Evolution II.* Oxford: Oxford University Press.

Hurford, J. (2014). *The Origins of Language: A Slim Guide.* Oxford: Oxford University Press.

Ilambu, O. (2001). "Ecology of Eastern Lowland Gorilla: Is There Enough Scientific Knowledge to Mitigate Conservation Threats Associated with Extreme Disturbances in Its Distribution Range?" In *The Apes: Challenges for the 21st Century*, 307–312. Conference proceedings. Brookfield, IL: Brookfield Zoo.

Irvine, E. (2016). "Method and Evidence: Gesture and Iconicity in the Evolution of Language." *Mind and Language* 31 (2): 221–247.

Isaac, B. (1987). "Throwing and Human Evolution." *African Archaeological Review* 5 (1): 3–17.

Jackendoff, R. (1999). "Possible Stages in the Evolution of the Language Capacity." *Trends in Cognitive Science* 3 (7): 272–279.

Jackendoff, R. (2007). *Language, Consciousness, Culture: Essays on Mental Structure.* Cambridge, MA: MIT Press.

Jackson, F. (1997). *From Metaphysics to Ethics.* Oxford: Oxford University Press.

Jackson, F. (2010). *Language, Names and Information.* Oxford: Wiley.

Jaeggi, A. V., and M. Gurven (2017). "Food-Sharing Models." In *The International Encyclopedia of Anthropology*, ed. H. Callan. New York: John Wiley.

Jeffares, B. (2014). "Back to *Australopithecus*: Utilizing New Theories of Cognition to Understand the Pliocene Hominins." *Biological Theory* 9 (1): 4–15.

Kahneman, D. (2011). *Thinking, Fast and Slow*. London: Macmillan.

Kaplan, H., S. Gangestad, et al. (2007). "The Evolution of Diet, Brain and Life History among Primates and Humans." In *Guts and Brains*, ed. W. Roebroeks, 47–90. Leiden: Leiden University Press.

Kaplan, H., K. Hill, et al. (2000). "A Theory of Human Life History Evolution: Diet, Intelligence and Longevity." *Evolutionary Anthropology* 9 (4): 156–185.

Kelly, D. (1984). "Archaeology of Aboriginal Fish Traps in the Murray-Darling Basin, Australia." Sydney: Charles Sturt University.

Kelly, L. (2015). *Knowledge and Power in Prehistoric Societies: Orality, Memory, and the Transmission of Culture*. Cambridge: Cambridge University Press.

Kelly, R. L. (2013). *The Lifeways of Hunter-Gatherers: The Foraging Spectrum*. Cambridge: Cambridge University Press.

Kelly, S. D., and R. B. Church (1997). "Can Children Detect Conceptual Information Conveyed through Other Children's Nonverbal Behaviors?" *Cognition and Instruction* 15 (1): 107–134.

Killin, A. (2017a). "Plio-Pleistocene Foundations of Musicality: The Co-Evolution of Hominin Cognition, Sociality and Music." *Biological Theory* 12 (4): 222–235.

Killin, A. (2017b). "Where Did Language Come From? Connecting Sign, Song, and Speech in Hominin Evolution." *Biology and Philosophy* 32 (6): 759–778.

Kirsh, D. (1995). "Complementary Strategies: Why We Use Our Hands When We Think." In *Proceedings of the Seventeenth Annual Conference of the Cognitive Science Society*, 212–217. Hillsdale, NJ: Lawrence Erlbaum.

Koechlin, E., and T. Jubault (2006). "Broca's Area and the Hierarchical Organization of Human Behavior." *Neuron* 50 (6): 963–974.

Kohn, M., and S. Mithen (1999). "Handaxes: Products of Sexual Selection?" *Antiquity* 73 (281): 518–526.

Krebs, J., and R. Dawkins (1984). "Animal Signals, Mind-Reading and Manipulation." In *Behavioural Ecology: An Evolutionary Approach*, ed. J. R. Krebs and N. B. Davies, 380–402. Oxford: Blackwell Scientific.

Krupenye, C., F. Kano, et al. (2016). "Great Apes Anticipate That Other Individuals Will Act According to False Beliefs." *Science* 354 (6308): 110–114.

Kuhn, S. (2020). *The Evolution of Paleolithic Technologies*. London: Routledge.

Laland, K. (2017). "The Origins of Language in Teaching." *Psychonomic Bulletin and Review* 24 (1): 225–231.

Layton, R., and S. O'Hara (2010). "Human Social Evolution: A Comparison of Hunter-Gatherer and Chimpanzee Social Organization." In *Social Brain, Distributed Mind*, ed. R. Dunbar, C. Gamble, and J. Gowlett, 83–113. Oxford: Oxford University Press.

Layton, R., S. O'Hara, and A. Bilsborough (2012). "Antiquity and Social Functions of Multilevel Social Organization among Human Hunter-Gatherers." *International Journal of Primatology* 33: 1215–1245.

Leavens, D., W. Hopkins, et al. (2005). "Understanding the Point of Chimpanzee Pointing: Epigenesis and Ecological Validity." *Current Directions in Psychological Science* 14 (4): 185–189.

Lee, R. B. (1979). *The !Kung San: Men, Women and Work in a Foraging Society*. Cambridge: Cambridge University Press.

Lehmann, J., A. H. Korstjens, et al. (2008). "Time Management in Great Apes: Implications for Gorilla Biogeography." *Evolutionary Ecology Research* 10 (4).

Levinson, S. C. (2013). "Recursion in Pragmatics." *Language* 89 (1): 149–162.

Levinson, S. (2016). "Turn-Taking in Human Communication: Origins and Implications for Language Processing." *Trends in Cognitive Sciences* 20 (1): 6–14.

Levinson, S., and D. Dediu (2018). "Neanderthal Language Revisited: Not Only Us." *Current Opinion in Behavioral Sciences* 21: 49–56.

Lewis, D. K. (1969). *Convention*. Cambridge, MA: Harvard University Press.

Lewis, J. (2009). "As Well as Words: Congo Pygmy Hunting, Mimicry, and Play." In *The Cradle of Language*, ed. R. Botha and C. Knight, 236–256. Oxford: Oxford University Press.

Lewis, J. (2015). "Where Goods Are Free but Knowledge Costs: Hunter-Gatherer Ritual Economics in Western Central Africa." *Hunter Gatherer Research* 1 (1).

Li, H., and R. Durbin (2011). "Inference of Human Population History from Individual Whole-Genome Sequences." *Nature* 475 (7357): 493.

Liebenberg, L. (1990). *The Art of Tracking and the Origin of Science*. Claremount, South Africa: David Philip.

Liebenberg, L. (2013). *The Origin of Science: On the Evolutionary Roots of Science and Its Implications for Self-Education and Citizen Science*. Cape Town: CyberTracker.

Lieberman, D., D. Bramble, et al. (2007). "The Evolution of Endurance Running and the Tyranny of Ethnography: A Reply to Pickering and Bunn." *Journal of Human Evolution* 53: 434–437.

Lieberman, P. (1998). *Eve Spoke: Human Language and Human Evolution*. New York: W. W. Norton.

Lieberman, P. (2012). "Vocal Tract Anatomy and the Neural Bases of Talking." *Journal of Phonetics* 40 (4): 608–622.

Lloyd, E. (2004). "Kanzi, Evolution, and Language." *Biology and Philosophy* 19: 577–588.

Lorenz, K. (1981). *The Foundations of Ethology*. New York: Springer.

Loulergue, L., A. Schilt, et al. (2008). "Orbital and Millennial-Scale Features of Atmospheric CH_4 over the Past 800,000 Years." *Nature* 453 (7193): 383–386.

Lupo, K. (2012). "On Early Hominin Meat Eating and Carcass Acquisition Strategies: Still Relevant after All These Years?" In *Stone Tools and Fossil Bones*, ed. M. Domínguez-Rodrigo, 115–151. New York: Cambridge University Press.

MacLarnon, A. M., and G. P. Hewitt (1999). "The Evolution of Human Speech: The Role of Enhanced Breathing Control." *American Journal of Physical Anthropology* 109 (3): 341–363.

Marean, C. W. (2010). "Pinnacle Point Cave 13B (Western Cape Province, South Africa) in Context: The Cape Floral Kingdom, Shellfish, and Modern Human Origins." *Journal of Human Evolution* 59 (3–4): 425–443.

Marean, C. W. (2011). "Coastal South Africa and the Coevolution of the Modern Human Lineage and the Coastal Adaptation." In *Trekking the Shore*, 421–440. New York: Springer.

Marean, C. W. (2014). "The Origins and Significance of Coastal Resource Use in Africa and Western Eurasia." *Journal of Human Evolution* 77: 17–40.

Marean, C. W. (2015). "An Evolutionary Anthropological Perspective on Modern Human Origins." *Annual Review of Anthropology* 44: 533–556.

Marean, C. W., M. Bar-Matthews, et al. (2007). "Early Human Use of Marine Resources and Pigment in South Africa during the Middle Pleistocene." *Nature* 449 (7164): 905–908.

Maricic, T., V. Günther, et al. (2013). "A Recent Evolutionary Change Affects a Regulatory Element in the Human FOXP2 Gene." *Molecular Biology and Evolution* 30 (4): 844–852.

Marlowe, F. W. (2010). *The Hadza: Hunter-Gatherers of Tanzania*. Berkeley: University of California Press.

Martínez, I., J. L. Arsuaga, et al. (2008). "Human Hyoid Bones from the Middle Pleistocene Site of the Sima de los Huesos (Sierra de Atapuerca, Spain)." *Journal of Human Evolution* 54 (1): 118–124.

Martínez, I., R. M. Quam, and M. Rosa, M. (2008). "Auditory Capacities of Human Fossils: A New Approach to the Origin of Speech." In *Proceedings of the 2nd ASA-EAA Joint Conference Acoustics*, 4177–4182.

Martínez, I., M. Rosa, et al. (2004). "Auditory Capacities in Middle Pleistocene Humans from the Sierra de Atapuerca in Spain." *Proceedings of the National Academy of Sciences* 101 (27): 9976–9981.

Martrat, B., O. Grimalt, et al. (2007). "Four Climate Cycles of Recurring Deep and Surface Water Destabilizations on the Iberian Margin." *Science* 317 (5837): 502–507.

Marwick, B. (2003). "Pleistocene Exchange Networks as Evidence for the Evolution of Language." *Cambridge Archaeological Journal* 13 (1): 67–81.

Maslin, M. (2017). *The Cradle of Humanity*. Oxford: Oxford University Press.

Maynard Smith, J., and D. Harper (2003). *Animal Signals*. Oxford: Oxford University Press.

Mazza, P. P. A., F. Martini, et al. (2006). "A New Palaeolithic Discovery: Tar-Hafted Stone Tools in a European Mid-Pleistocene Bone-Bearing Bed." *Journal of Archaeological Science* 33 (9): 1310–1318.

McBrearty, S., and A. Brooks (2000). "The Revolution That Wasn't: A New Interpretation of the Origin of Modern Human Behavior." *Journal of Human Evolution* 39 (5): 453–563.

McNeill, D. (2008). *Gesture and Thought*. Chicago: University of Chicago Press.

McNeill, D., J. Cassell, and K. E. McCullough (1994). "Communicative Effects of Speech-Mismatched Gestures." *Research on Language and Social Interaction* 27 (3): 223–237.

McPherron, S. P., Z. Alemseged, et al. (2010). "Evidence for Stone-Tool-Assisted Consumption of Animal Tissues before 3.39 Million Years Ago at Dikika, Ethiopia." *Nature* 466: 857–860.

Meggitt, M. (1965). *Desert People: A Study of the Walbiri Aborigines of Central Australia*. Sydney: Angus and Robinson.

Meir, I., W. Sandler, et al. (2010). "Emerging Sign Languages." In *Oxford Handbook of Deaf Studies, Language, and Education*, ed. M. Marschark, 267–280. Oxford: Oxford University Press.

Meloy, J. R. (2006). "Empirical Basis and Forensic Application of Affective and Predatory Violence." *Australian and New Zealand Journal of Psychiatry* 40 (6–7): 539–547.

Mercader, J., H. Barton, et al. (2007). "4,300-Year-Old Chimpanzee Sites and the Origins of Percussive Stone Technology." *Proceedings of the National Academy of Sciences* 104 (9): 3043–3048.

Mercader, J., M. Panger, and C. Boesch (2002). "Excavation of a Chimpanzee Stone Tool Site in the African Rainforest." *Science* 296 (5572): 1452–1455.

Mercier, H., and D. Sperber (2017). *The Enigma of Reason*. Cambridge, MA: Harvard University Press.

Milks, A., D. Parker, et al. (2019). "External Ballistics of Pleistocene Hand-Thrown Spears: Experimental Performance Data and Implications for Human Evolution." *Scientific Reports* 9 (820).

Mirazón Lahr, M., and R. A. Foley (1998). "Towards a Theory of Modern Human Origins: Geography, Demography, and Diversity in Recent Human Evolution." *American Journal of Physical Anthropology* 107 (S27): 137–176.

Mithen, S. (1996). *The Prehistory of the Mind*. London: Phoenix Books.

Mithen, S. (2005). *The Singing Neanderthals: The Origins of Music, Language, Mind and Body*. London: Weidenfeld & Nicholson.

Mithen, S. (2009). "Holistic Communication and the Coevolution of Language and Music: Resurrecting an Old Idea." In *The Prehistory of Language*, ed. R. Botha and C. Knight, 58–76. Oxford: Oxford University Press.

Morphy, H. (1989). "On Representing Ancestral Beings." In *Animals into Art*, 144–160. London: Routledge.

Morphy, H. (1991). *Ancestral Connections: Art and an Aboriginal System of Knowledge*. Chicago: University of Chicago Press.

Murgatroyd, S. (2002). *The Dig Tree: The Story of Bravery, Insanity, and the Race to Discover Australia's Wild Frontier*. Melbourne: Text Classics.

Mussi, M. (2007). "Women of the Middle Latitudes: The Earliest Peopling of Europe from a Female Perspective." In *Guts and Brains*, ed. W. Roebroeks, 168–183. Leiden: Leiden University Press.

Nevins, A., D. Pesetsky, et al. (2009). "Pirahã Exceptionality: A Reassessment." *Language* 85 (2): 355–404.

Nishida, A. K., N. Nordi, and R. R. Alves (2006a). "The Lunar-Tide Cycle Viewed by Crustacean and Mollusc Gatherers in the State of Paraíba, Northeast Brazil and Their Influence in Collection Attitudes." *Journal of Ethnobiology and Ethnomedicine* 2 (1): 1.

Nishida, A. K., N. Nordi, and R. R. Alves (2006b). "Molluscs Production Associated to Lunar Tide Cycle: A Case Study in Paraíba State under Ethnoecology Viewpoint." *Journal of Ethnobiology and Ethnomedicine* 2 (1): 28.

O'Driscoll, C., and J. Thompson (2018). "The Origins and Early Elaboration of Projectile Technology." *Evolutionary Anthropology* 27: 30–45.

Ofek, H. (2001). *Second Nature: Economic Origins of Human Evolution.* Cambridge: Cambridge University Press.

Parker, C., E. Keefe, et al. (2016). "The Pyrophilic Primate Hypothesis." *Evolutionary Anthropology* 25: 54–63.

Patterson, D. B., D. R. Braun, et al. (2019). "Comparative Isotopic Evidence from East Turkana Supports a Dietary Shift within the Genus *Homo*." *Nature Ecology and Evolution* 3: 1048–1056.

Pearce, E., J. Launay, and R. I. Dunbar (2015). "The Ice-Breaker Effect: Singing Mediates Fast Social Bonding." *Royal Society Open Science* 2 (10): 150221.

Pearce, E., C. Stringer, and R. I. Dunbar (2013). "New Insights into Differences in Brain Organization between Neanderthals and Anatomically Modern Humans." *Proceedings of the Royal Society B: Biological Sciences* 280 (1758): 20130168.

Perelman, P., W. E. Johnson, et al. (2011). "A Molecular Phylogeny of Living Primates." *PLoS Genetics* 7 (3): e1001342.

Petraglia, M. D., C. Shipton, and K. Paddayya (2005). "Life and Mind in the Acheulean." In *The Hominid Individual in Context: Archaeological Investigations of Lower and Middle Palaeolithic Landscapes, Locales and Artefacts*, ed. C. Gamble and M. Porr, 197–219. London: Routledge.

Pettit, Pa. (2011). *The Palaeolithic Origins of Human Burial.* London: Routledge.

Pettit, Ph. (1993). *The Common Mind: An Essay on Psychology, Society, and Politics.* Oxford: Oxford University Press.

Pickering, T. R. (2013). *Rough and Tumble: Aggression, Hunting, and Human Evolution.* Los Angles: University of California Press.

Pickering, T. R., and H. Bunn (2007). "The Endurance Running Hypothesis and Hunting and Scavenging in Savanna–Woodlands." *Journal of Human Evolution* 53: 438–442.

Pickering, T. R., and H. Bunn (2012). "Meat Foraging by Pleistocene African Hominins: Tracking Behavioral Evolution beyond Baseline Inferences of Early Access to Carcasses." In *Stone Tools and Fossil Bones*, ed. M. Domínguez-Rodrigo, 152–173. New York: Cambridge University Press.

Pickering, T. R., and M. Domínguez-Rodrigo (2012). "Can We Use Chimpanzee Behavior to Model Early Hominin Hunting?" In *Stone Tools and Fossil Bones*, ed. M. Domínguez-Rodrigo, 174–203. New York: Cambridge University Press.

Pilley, J. W. (2013). *Chaser: Unlocking the Genius of the Dog Who Knows a Thousand Words*. Boston: Houghton Mifflin Harcourt.

Pinker, S. (2011). *The Better Angels of Our Nature.* New York: Viking.

Planer, R. J. (2017a). "Protolanguage Might Have Evolved before Ostensive Communication." *Biological Theory* 12 (2): 72–84.

Planer, R. J. (2017b). "Talking about Tools: Did Early Pleistocene Hominins Have a Protolanguage?" *Biological Theory* 12 (4): 211–221.

Planer, R. J. (2020a). "Towards an Evolutionary Account of Human Kinship Systems." *Biological Theory*, 1–14.

Planer, R. J. (2020b). Review of Richard Wrangham's "The Goodness Paradox." *British Journal Review of Books*.

Planer, R. J. (2021). "Theory of Mind, System-2 Thinking, and the Origins of Language." In *Explorations in Archaeology and Philosophy*, ed. A. Killin and S. Allen-Hermanson. Synthese Library. Cham, Switzerland: Springer.

Planer, R. J., and D. Kalkman (2019). "Arbitrary Signals and Cognitive Complexity." *British Journal for the Philosophy of Science*.

Planer, R. J., and P. Godfrey-Smith (forthcoming). "Communication and Representation Understood as Sender–Receiver Coordination." *Mind and Language*.

Ploog, D. (2002). "Is the Neural Basis of Vocalization Different in Non-human Primates and *Homo sapiens*?" In *The Speciation of Modern Homo sapiens*, ed. T. J. Crow. Oxford: Oxford University Press.

Potts, R. (1996). *Humanity's Descent: The Consequences of Ecological Instability*. New York: Avon.

Potts, R. (1998). "Variability Selection in Hominid Evolution." *Evolutionary Anthropology* 7 (3): 81–96.

Potts, R., and J. T. Faith (2015). "Alternating High and Low Climate Variability: The Context of Natural Selection and Speciation in Plio-Pleistocene Hominin Evolution." *Journal of Human Evolution* 87: 5–20.

Powell, A., S. Shennan, et al. (2009). "Late Pleistocene Demography and the Appearance of Modern Human Behavior." *Science* 324(June 5): 298–1301.

Progovac, L. (2015). *Evolutionary Syntax*. Oxford: Oxford University Press.

Provine, R. R. (2001). *Laughter: A Scientific Investigation*. New York: Penguin.

Pruetz, J. D., and P. Bertolani (2009). "Chimpanzee (*Pan troglodytes verus*) Behavioral Responses to Stresses Associated with Living in a Savanna-Mosaic Environment: Implications for Hominin Adaptations to Open Habitats." *PaleoAnthropology*, 252.

Pruetz, J. D., and N. M. Herzog (2017). "Savanna Chimpanzees at Fongoli, Senegal, Navigate a Fire Landscape." *Current Anthropology* 58 (S16): S337–S350.

Prüfer, K., F. Racimo, et al. (2014). "The Complete Genome Sequence of a Neanderthal from the Altai Mountains." *Nature* 505 (7481): 43–49.

Railton, P. (2006). "Normative Guidance." *Oxford Studies in Metaethics* 1 (3): 2–33.

Rauscher, F. H., R. M. Krauss, and Y. Chen (1996). "Gesture, Speech, and Lexical Access: The Role of Lexical Movements in Speech Production." *Psychological Science* 7 (4): 226–231.

Richerson, P. J., and R. Boyd (2005). *Not by Genes Alone: How Culture Transformed Human Evolution.* Chicago: University of Chicago Press.

Richerson, P., and R. Boyd (2013). "Rethinking Paleoanthropology: A World Queerer Than We Had Supposed." In *The Evolution of Brain, Mind, and Culture,* ed. G. Hatfield and H. Pittman. Philadelphia: University of Pennsylvania Press.

Richman, B. (1993). "On the Evolution of Speech: Singing as the Middle Term." *Current Anthropology* 34 (5): 721–722.

Ridley, M. (1986). *Evolution and Classification: The Reformulation of Cladism.* London: Longman.

Rizzolatti, G., L. Fadiga, et al. (1996). "Premotor Cortex and the Recognition of Motor Actions." *Cognitive Brain Research* 3 (2): 131–141.

Roach, N. T., and B. G. Richmond (2015). "Clavicle Length, Throwing Performance and the Reconstruction of the *Homo erectus* Shoulder." *Journal of Human Evolution* 80: 107–113.

Roberts, P., and B. Stewart (2018). "Defining the 'Generalist Specialist' Niche for Pleistocene *Homo sapiens.*" *Nature Human Behaviour* 2: 542–550.

Rossano, M. J. (2010). "Making Friends, Making Tools, and Making Symbols." *Current Anthropology* 51 (S1): S89–S98.

Rowson, B., B. H. Warren, and C. F. Ngereza (2010). "Terrestrial Molluscs of Pemba Island, Zanzibar, Tanzania, and Its Status as an 'Oceanic' Island." *ZooKeys* 70: 1–39.

Russon, A., and K. Andrews (2011). "Pantomime in Great Apes: Evidence and Implications." *Communicative and Integrative Biology* 4 (3): 315–317.

Russon, A., and K. Andrews (2015). "Orangutan Pantomime: Elaborating the Message." *Biology Letters* 1–4.

Saffran, J., M. Hauser, et al. (2008). "Grammatical Pattern Learning by Human Infants and Cotton-Top Tamarin Monkeys." *Cognition* 107 (2): 479–500.

Sahle, Y., W. K. Hutchings, et al. (2013). "Earliest Stone-Tipped Projectiles from the Ethiopian Rift Date to >279,000 Years Ago." *PLoS One* 8 (11): e78092.

Salali, L., R. Mace, et al. (2017). "Cooperation and the Evolution of Hunter-Gatherer Storytelling." *Nature Communications* 8 (1): 1853.

Sandgathe, D., H. Dibble, et al. (2011). "On the Role of Fire in Neandertal Adaptations in Western Europe: Evidence from Pech de l'Azé IV and Roc de Marsal, France." *PaleoAnthropology* 2011: 216–224.

Satterthwait, L. (1986). "Aboriginal Australian Net Hunting." *Mankind* 16 (1): 31–48.

Satterthwait, L. (1987). "Socioeconomic Implications of Australian Aboriginal Net Hunting." *Man* 22 (4): 613–636.

Schel, A. M., S. W. Townsend, et al. (2013). "Chimpanzee Alarm Call Production Meets Key Criteria for Intentionality." *PLoS One* 8 (10): e76674.

Schiffels, S., and R. Durbin (2014). "Inferring Human Population Size and Separation History from Multiple Genome Sequences." *Nature Genetics* 46 (8): 919–925.

Scott-Phillips, T. (2015). *Speaking Our Minds*. London: Palgrave-Macmillan.

Scott-Phillips, T. C., and R. A. Blythe (2013). "Why Is Combinatorial Communication Rare in the Natural World, and Why Is Language an Exception to This Trend?" *Journal of the Royal Society Interface* 10.

Seabright, P. (2010). *The Company of Strangers: A Natural History of Economic Life*. Princeton: Princeton University Press.

Searcy, W., and S. Nowicki (2005). *The Evolution of Animal Communication: Reliability and Deception in Signaling Systems*. Princeton: Princeton University Press.

Segal, N. L. (2012). *Born Together—Reared Apart: The Landmark Minnesota Twin Study*. Cambridge, MA: Harvard University Press.

Seyfarth, R., and D. Cheney (2010). "Production, Usage, and Comprehension in Animal Vocalizations." *Brain and Language* 115: 92–100.

Seyfarth, R., and D. Cheney (2014). "The Evolution of Language from Social Cognition." *Current Opinion in Neurobiology* 28: 5–9.

Seyfarth, R., and D. Cheney (2018a). "Pragmatic Flexibility in Primate Vocal Production." *Current Opinion in Behavioral Sciences* 21: 56–61.

Seyfarth, R., and D. Cheney (2018b). *The Social Origins of Language*. Princeton: Princeton University Press.

Shahack-Gross, R., F. Berna, et al. (2014). "Evidence for the Repeated Use of a Central Hearth at Middle Pleistocene (300 ky Ago) Qesem Cave, Israel." *Journal of Archaeological Science* 44: 12–21.

Shaw-Williams, K. (2014). "The Social Trackways Theory of the Evolution of Human Cognition." *Biological Theory* 9 (1): 16–26.

Shaw-Williams, K. (2017). "The Social Trackways Theory of the Evolution of Language." *Biological Theory* 12 (4): 195–210.

Shea, J. J. (2006). "The Origins of Lithic Projectile Point Technology: Evidence from Africa, the Levant, and Europe." *Journal of Archaeological Science* 33 (6): 823–846.

Shea, J. J., and M. L. Sisk (2010). "Complex Projectile Technology and *Homo sapiens* Dispersal into Western Eurasia." *PaleoAnthropology* 2010: 100–122.

Shipton, C. (2019). "The Evolution of Social Transmission in the Acheulean." In *Squeezing Minds from Stones: Cognitive Archaeology and the Evolution of the Human Mind.* ed. K. Overmann and F. Coolidge. Oxford: Oxford University Press.

Simões-Costa, M., and M. E. Bronner (2015). "Establishing Neural Crest Identity: A Gene Regulatory Recipe." *Development* 142 (2): 242–257.

Skyrms, B. (2010). *Signals: Evolution, Learning, and Information.* New York: Oxford University Press.

Slocombe, K., T. Kaller, et al. (2010). "Production of Food-Associated Calls in Wild Male Chimpanzees Is Dependent on the Composition of the Audience." *Behavioral Ecology and Sociobiology* 64: 1959–1966.

Smith, D., P. Schlaepfer, et al. (2017). "Cooperation and the Evolution of Hunter-Gatherer Storytelling." *Nature Communications* 8 (1): 1–9.

Spencer, B. (1928). *Wanderings in Wild Australia.* London: Macmillan.

Sperber, D., and D. Wilson (1986). *Relevance: Communication and Cognition.* Oxford: Blackwell.

Stamps, J. A., and M. Buechner (1985). "The Territorial Defense Hypothesis and the Ecology of Insular Vertebrates." *Quarterly Review of Biology* 60 (2): 155–181.

Stanford, C. (2018). *The New Chimpanzee: A Twenty-First Century Portrait of Our Closest Kin.* Cambridge, MA: Harvard University Press.

Sterelny, K. (2011). "From Hominins to Humans: How *sapiens* Became Behaviourally Modern." *Philosophical Transactions of the Royal Society B* 366 (1566): 809–822.

Sterelny, K. (2012a). *The Evolved Apprentice* Cambridge, MA: MIT Press.

Sterelny, K. (2012b). "Language, Gesture, Skill: The Co-evolutionary Foundations of Language." *Philosophical Transactions of the Royal Society B* 367: 2141–2151.

Sterelny, K. (2014). "A Paleolithic Reciprocation Crisis: Symbols, Signals, and Norms." *Biological Theory* 9 (1): 65–77.

Sterelny, K. (2016a). "Cumulative Cultural Evolution and the Origins of Language." *Biological Theory* 11 (3): 173–186.

Sterelny, K. (2016b). "Deacon's Challenge: From Calls to Words." *Topoi* 35 (1): 271–282.

Sterelny, K. (2017). "Cultural Evolution in California and Paris." *Studies in History and Philosophy of Science Part C* 62: 42–50.

Sterelny, K. (2019a). "Norms and Their Evolution." In *Handbook of Cognitive Archaeology: Psychology in Prehistory*, ed. T. Henley, M. Rossano, and E. P. Kardas, 375–397. New York: Routledge.

Sterelny, K. (2019b). "The Origins of Multi-Level Society." *Topoi.*

Sterelny, K. (2020). "Demography and Cultural Complexity." *Synthese.*

Sterelny, K. (2021). *The Pleistocene Social Contract: Culture and Cooperation in the Hominin Lineage.* New York: Oxford University Press.

Stiner, M. C. (2002). "Carnivory, Coevolution, and the Geographic Spread of the Genus *Homo.*" *Journal of Archaeological Research* 10 (1): 1–63.

Stiner, M. (2013). "An Unshakable Middle Paleolithic? Trends versus Conservatism in the Predatory Niche and Their Social Ramifications." *Current Anthropology* 54 (S8): S288–S304.

Stiner, M., A. Gopher, et al. (2011). "Hearth-Side Socioeconomics, Hunting and Paleoecology during the Late Lower Paleolithic at Qesem Cave, Israel." *Journal of Human Evolution* 60: 213–233.

Stoessel, A., R. David, P. Gunz, T. Schmidt, F. Spoor, and J. J. Hublin (2016). "Morphology and Function of Neandertal and Modern Human Ear Ossicles." *Proceedings of the National Academy of Sciences* 113 (41): 11489–11494.

Stout, D. (2011). "Stone Toolmaking and the Evolution of Human Culture and Cognition." *Philosophical Transactions of the Royal Society B* 366: 1050–1059.

Stout, D., and T. Chaminade (2009). "Making Tools and Making Sense: Complex, Intentional Behaviour in Human Evolution." *Cambridge Archaeological Journal* 19 (1): 85–96.

Stout, D., and T. Chaminade (2012). "Stone Tools, Language and the Brain in Human Evolution." *Philosophical Transactions of the Royal Society B* 367 (1585): 75–87.

Striedter, G. F. (2005). *Principles of Brain Evolution.* Sunderland, MA: Sinauer Associates.

Sugiyama, M. (2001). "Food, Foragers and Folklore: The Role of Narrative in Human Subsistence." *Evolution and Human Behavior* 22 (4): 221–240.

Sykes, R. W. (2015). "To See a World in a Hafted Tool: Birch Pitch Composite Technology, Cognition and Memory in Neanderthals." In *Settlement, Society and Cognition in Human Evolution: Landscapes in the Mind*, ed. F. Coward, R. Hosfield, et al., 117–137. Cambridge: Cambridge University Press.

Tattersall, I. (2016). "Language Origins: An Evolutionary Framework." *Topoi* 37: 289–296.

Tennie, C., D. R. Braun, et al. (2016). "The Island Test for Cumulative Culture in Paleolithic Cultures." In *The Nature of Culture*, ed. M. N. Haidle, N. Conard, and M. Bolus, 121–133. Dordrecht: Springer.

Tennie, C., J. Call, et al. (2009). "Ratcheting Up the Ratchet: On the Evolution of Cumulative Culture." *Philosophical Transactions of the Royal Society B* 364: 2405–2415.

Tennie, C., L. Premo, et al. (2017). "Early Stone Tools and Cultural Transmission: Resetting the Null Hypothesis, with Commentaries and a Response." *Current Anthropology* 58 (5): 652–672.

Theofanopoulou, C., S. Gastaldon, et al. (2017). "Comparative Genomic Evidence for Self-Domestication in *Homo sapiens*." *bioRxiv*, 125799.

Thieme, H. (1997). "Lower Palaeolithic Hunting Spears from Germany." *Nature* 385 (27 February): 807–810.

Thompson, J., S. Carvalho, et al. (2019). "Origins of the Human Predatory Pattern: The Transition to Large Animal Exploitation by Early Hominins." *Current Anthropology* 60: 1–23.

Thompson, L. A., and D. W. Massaro (1994). "Children's Integration of Speech and Pointing Gestures in Comprehension." *Journal of Experimental Child Psychology* 57 (3): 327–354.

Thornton, A., and N. Raihani (2008). "The Evolution of Teaching." *Animal Behaviour* 75 (6): 1823–1836.

Tomasello, M. (2003). *Constructing a Language: A Usage-Based Theory of Language Acquisition*. Cambridge, MA: Harvard University Press.

Tomasello, M. (2007). "If They Are So Good at Language, Then Why Don't They Talk? Hints from Apes' and Humans' Uses of Gesture." *Language Learning and Development* 3 (2): 133–156.

Tomasello, M. (2008). *Origins of Human Communication*. Cambridge, MA: MIT Press.

Tomasello, M. (2009). *Why We Cooperate*. Cambridge, MA: MIT Press.

Tomasello, M. (2014). *A Natural History of Human Thinking*. Cambridge, MA: Harvard University Press.

Tomasello, M. (2016). *A Natural History of Human Morality*. Cambridge, MA: Harvard University Press.

Tomasello, M., A. P. Melis, et al. (2012). "Two Key Steps in the Evolution of Human Cooperation: The Interdependence Hypothesis." *Current Anthropology* 53 (6): 673–692.

Toth, N., and K. Schlick (2019). "Why Did the Acheulean Happen? Experimental Studies into the Manufacture and Function of Acheulean Artifacts." *L'anthropologie* 123: 724–768.

Tramacere, A., and R. Moore (2017). "Reconsidering the Role of Manual Imitation in Language Evolution." *Topoi* 37 (2): 319–328.

Truswell, R. (2017). "Dendrophobia in Bonobo Comprehension of Spoken English." *Mind and Language* 32 (4): 395–415.

Trut, L., I. Oskina, and A. Kharlamova (2009). "Animal Evolution during Domestication: The Domesticated Fox as a Model." *Bioessays* 31 (3): 349–360.

Twomey, T. (2013). "The Cognitive Implications of Controlled Fire Use by Early Humans." *Cambridge Archaeological Journal* 23 (1): 113–128.

Twomey, T. (2014). "How Domesticating Fire Facilitated the Evolution of Human Cooperation." *Biology and Philosophy* 29 (1): 89–99.

Veroude, K., Y. Zhang-James, et al. (2016). "Genetics of Aggressive Behavior: An Overview." *American Journal of Medical Genetics Part B: Neuropsychiatric Genetics* 171 (1): 3–43.

Villamil, C. I. (2014). "An Analysis of *Homo erectus* Vertebral Canal Morphology and Its Relationship to Vertebral Formula Variation in Recent Humans." *American Journal of Physical Anthropology* 153: 261–261.

Walker, R. S., M. Flinn, et al. (2010). "Evolutionary History of Partible Paternity in Lowland South America." *Proceedings of the National Academy of Sciences* 107 (45): 19195–19200.

Wallace, A. R. (1891). *Natural Selection and Tropical Nature: Essays on Descriptive and Theoretical Biology*. London: Macmillan.

Warneken, F., and M. Tomasello (2006). "Altruistic Helping in Human Infants and Young Chimpanzees." *Science* 311 (3 March): 1301–1303.

Watts, D. P. (1988). "Environmental Influences on Mountain Gorilla Time Budgets." *American Journal of Primatology* 15 (3): 195–211.

Weinstein, D., J. Launay, et al. (2016). "Group Music Performance Causes Elevated Pain Thresholds and Social Bonding in Small and Large Groups of Singers." *Evolution and Human Behavior* 37 (2): 152–158.

West-Eberhard, M. J. (2003). *Developmental Plasticity and Evolution*. Oxford: Oxford University Press.

Whallon, R. (2011). "An Introduction to Information and Its Role in Hunter-Gatherer Bands." In *Information and Its Role in Hunter-Gatherer Bands*, ed. R. Whallon, W. A. Lovis, and R. Hitchcock, 1–28. Los Angeles: UCLA/Cotsen Institute of Archaeology Press.

Wiessner, P. (2002). "Hunting, Healing, and hxaro Exchange: A Long-Term Perspective on !Kung (Ju/'hoansi) Large-Game Hunting." *Evolution and Human Behavior* 23 (6): 407–436.

Wiessner, P. W. (2014). "Embers of Society: Firelight Talk among the Ju/'hoansi Bushmen." *Proceedings of the National Academy of Sciences* 111 (39): 14027–14035.

Wilkins, D. P. (1997). "Alternative Representations of Space: Arrernte Narratives in Sand." In *The Visual Narrative Reader*, ed. N. Cohn, 252–281. London: Bloomsbury.

Wilkins, J., and M. Chazan (2012). "Blade Production 500 Thousand Years Ago at Kathu Pan 1, South Africa: Support for a Multiple Origins Hypothesis for Early Middle Pleistocene Blade Technologies." *Journal of Archaeological Science* 39: 1883–1900.

Wilkins, J., B. J. Schoville, et al. (2012). "Evidence for Early Hafted Hunting Technology." *Science* 338: 942–945.

Wrangham, R. W. (1977). "Feeding Behaviour of Chimpanzees in Gombe National Park, Tanzania." In *Primate Ecology: Studies of Feeding and Ranging Behaviour in Lemurs, Monkeys and Apes*, 504–538. New York: Academic Press.

Wrangham, R. (2017). "Control of Fire in the Paleolithic: Evaluating the Cooking Hypothesis." *Current Anthropology* 58 (S16): S303–S313.

Wrangham, R. (2018). *The Goodness Paradox*. Cambridge, MA: Harvard University Press.

Wrangham, R. W. (2019). "Hypotheses for the Evolution of Reduced Reactive Aggression in the Context of Human Self-Domestication." *Frontiers in Psychology* 10.

Wray, A. (1998). "Protolanguage as a Holistic System for Social Interaction." *Language and Communication* 18 (1): 47–67.

Wray, A. (2002). "Dual Processing in Protolanguage: Performance without Competence." In Wray, *The Transition to Language*, 113–137. Oxford: Oxford University Press.

Wray, A. (2005). "The Explanatory Advantages of the Holistic Protolanguage Model: The Case of Linguistic Irregularity. Commentary on Arbib." *Behavioral and Brain Sciences* 28 (2): 147–148.

Zachos, J., M. Pagani, et al. (2001). "Trends, Rhythms, and Aberrations in Global Climate 65 Ma to Present." *Science* 292 (5517): 686–693.

Zawidzki, T. W. (2013). *Mindshaping: A New Framework for Understanding Human Social Cognition*. Cambridge, MA: MIT Press.

Index